W9-BSX-475

COUNTRY GRIT

COUNTRY GRIT

A FARMOIR OF FINDING PURPOSE AND LOVE

SCOTTIE JONES

Skyhorse Publishing

Skyhorse Publishing books may be purchased in bulk at special discounts for sales promotion, corporate gifts, fund-raising, or educational purposes. Special editions can also be created to specifications. For details, contact the Special Sales Department, Skyhorse Publishing, 307 West 36th Street, 11th Floor, New York, NY 10018 or info@skyhorsepublishing.com.

Visit our website at www.skyhorsepublishing.com.

10 9 8 7 6 5 4 3 2 1

Library of Congress Cataloging-in-Publication Data

Names: Jones, Scottie Brown, author.
Title: Country grit : a farmoir of finding purpose and love / Scottie Brown Jones.
Description: New York, NY : Skyhorse Publishing, [2017]
Identifiers: LCCN 2017015225 (print) | LCCN 2017018539 (ebook) | ISBN 9781510722156 (ebook) | ISBN 9781510722149 (hardcover : alk. paper)
Subjects: LCSH: Farmers--United States--Biography. | Autobiography.
Classification: LCC S417.J66 (ebook) | LCC S417.J66 J66 2017 (print) | DDC
630.92--dc23
LC record available at https://lccn.loc.gov/2017015225

Cover design by Erin Seaward-Hiatt
Cover photo by Kristi Crawford
Author photo by Shawn Linehan

Printed in the United States of America

CONTENTS

Introduction 🐑 *ix*

PART ONE 🐑 1

PART TWO 🐑 49

PART THREE 🐑 83

PART FOUR 🐑 117

PART FIVE 🐑 157

Epilogue: Stay on the Farm 🐑 *195*
Addendum: A Primer on Farming 🐑 *199*
Acknowledgments 🐑 *215*

For Caitlin and Annie, who may have questioned our sanity at moving to a farm but not the adventure.

INTRODUCTION

This book began as a series of letters to friends back home. Well, some would call them letters; anguished cries for help might be another interpretation. My friends found my pain completely amusing and suggested that I turn the letters into a blog, which then morphed into a book. That should explain why this reads like a crossbred blog-book.

Country Grit is constructed around a series of vignettes that are absolutely true—if we're speaking of my experience. If we're speaking of objective facts, you should know all the characters are composites. Please don't come to my little town and expect to meet any of the characters in this book, because they don't exist. That's why I say the experiences I report on are real, just not the people. The real people in the real town I live in value their privacy and don't take kindly to strangers poking at them.

While researching this book, I found there are hundreds of farm memoirs. So many, I felt it deserved its own genre, hence the term: farmoir. It amazed me how similar the experiences were in each of these books. I considered that someone, possibly one of my siblings who were always jealous of my demure charm, leaked my letters, leading to hundreds of cases of plagiarism. Except the farmoirs go back nearly a century, just about the time Americans began moving off the farm and into cities in large numbers. So besides the funny animal stories, there is something more going on here, some nostalgia for what was left behind.

Urban life leaves us with a hunger for the rural life and its connection with the natural world. This hunger encourages romantic visions of rural life leading to a kind of delusional thinking. My farmoir chronicles the undoing of those delusions, including the destructive consequences that follow whenever we ignore reality. Anyone who has started a new enterprise or made a radical change in their life will find parallels in my story. It always starts with a dream of how we hope our new life will be, and then reality intervenes. We are challenged, and as a consequence we are changed. While there are similarities shared by any new undertaking, there is something special about farming that lends itself to romanticism, and an extra degree of delusional thinking, that brings this process into sharper focus. See if you agree.

The reality of farming is business. Throughout the book I offer insights into the economics that drive the business of farming. Admittedly it's pretty dry stuff, but hugely important since the business of farming ensures that you will eat today and tomorrow. The wealth of a country is built on its farming and especially on the efficiency of its farmers. It touches all of us every day, and so it's good citizenship to know something about the business of our food supply. Frankly, I'm surprised it isn't part of a standard high school education. For those who really want to wonk out on agricultural policy (and God bless you if you're one of them because we need you), I've included a little addendum. I owe an immense debt to the real scholars: Paul

Conkin (*A Revolution Down On The Farm*); David Danbom (*Born In The Country*); and Daniel Imhoff (*Food Fight*).

A number of friends lent their assistance to this project, including: Max and Dave Hanson, Karen and Allan Six, Craig Zaffaroni, Bert Banton, Linda and Ken Worley, Matt and John Clark, Mary and Steve O'Brien, Tanya Freeman, Nancy and Paul Cooke, Carolyn Avery, Russ Kaufman, Larry and Shirley Cole, Chuck and Lisa Smith, Mike and Liz Behrenfeld, Janet and Rolfe Hagen, and Curtis Koenig.

My other debt is to my family, whose patience and support made this possible. I'm speaking of my two daughters, Caitlin and Annie, but most importantly, my husband and collaborator, Greg, who provided extensive writing, research, and editorial assistance in this undertaking. And, of course, my grandson, Henry.

And with that, here's my story, such as it is.

PART ONE

THE BEGINNING

Let me begin with a simple truth: I was not born to be a farmer.

I was born to the manicured lawns of suburban Connecticut. My father, in his daily commute to New York City, never dreamed of his little girl knee-deep in mud, wrestling sheep for a living. The years of boarding school and the advanced degree in art history did not prepare me for the agricultural life. I am presently a farmer by dint of circumstance—call it a karmic accident. Someone threw a switch and my life went barreling down an entirely different track. And that someone was my loving spouse—might as well name names. Call it karmic because I chose him. Call it accident because . . . well, I'm getting to that part.

Men are itchy creatures—and never more itchy than when life is most content. So, as my husband was reaching the zenith of his career as a psychologist he became the most discontent. Every aspect of our life together was a little too much or not quite enough for him, as he searched for targets that would reflect his discontent. I took to walking around him with my invisibility shield up, which wasn't hard. Our lives were defined by well-rehearsed roles that kept us invisible to each other most of the time.

In fairness, my husband had a lot of help with his discontent. It was not an easy time to be a practicing psychologist. The crushing weight of managed health "cost" was turning sensitive caregivers into beleaguered accountants. Hours on the phone arguing care with insurance gatekeepers arguing costs left him deflated. He would say the profession he had dedicated his life to, and found so rewarding, had changed beyond his recognition. I would say men are itchy creatures.

For my part, I was happy in my work managing the retail end of a large urban zoo. I was responsible for a diverse set of income streams: the gate, gift shops, concessions, special events, and such. Each day brought new challenges—financial pinches, marketing glitches, and mechanical breakdowns. And each problem came with a disparate set of personalities to bridge. When it became too much, there were always the animals. Just across the fence, the animals' uncomplicated presence lent a grounding perspective to whatever problem was vexing me at the moment. Looking back, I was already in training for my next career. Mechanical breakdowns, teeth-grinding finances, marketing, and the animals—it's all there in farming, with a lot fewer resources to get the job done. After all, it takes a lot of people to run a zoo.

Our home life was dominated by our work life. It wasn't exactly bad, just compartmentalized. We were two high-functioning professionals in the urban-industrial, twenty-first century American experience. Both our children had recently matriculated to college, leaving us more time to pursue our careers. The models of efficiency we utilized at the office carried over to our chores, hobbies, and friendships, meaning we were very productive and rather affluent. We were also leading separate lives—separated by the compartments we created and the efficiency we expected. Lives that left us feeling vaguely lonely and disconnected in a multitude of ways that were hard to define without appearing whiney or ungrateful.

The next part of the karmic collision was contributed by the city itself. Rising from the desert, Phoenix was the source of plentiful jobs, cheap housing, and epic sprawl. There are no easy commutes in Phoenix, which is to say, there are no good starts to the day. For a third of the year the temperature exceeds 100 degrees. Commuting in that advanced state of swelter involves sitting in a carbon-belching, coffee-stained, bank-owned, all-terrain-cubicle, in eight lanes of gridlock, with direct sun baking you like truck-stop beef jerky. It follows that your thoughts turn to places you'd rather be—places that are cooler . . . greener . . . wetter. Please, God, a little wetter, with that faint promise that life can renew.

The final part really was inevitable. The other car crossed the line, demolishing our Acura, almost demolishing my husband, and ruining the day's commute. He survived but lost the use of his left hand. There would be months of rehabilitation. That meant months with little else to do but ponder all the parts that itch. He began an online affair with real estate, staying up late, staring at the computer screen, and conjuring a new life. Once conjured, he began to fact-check. Being a diligent researcher, this went on for months. He called it testing the hypothesis to reach an informed decision. I called it scratching the itch.

For my husband, the violence of that terrible day stripped away his "suburban pretense." It was not more things we needed, not more "stuff," but more of a connection to life itself. We needed to live more simply—in harmony with nature. We needed to get back to the land. We needed to get back to each other.

I had to admit, I kind of liked the sound of that last part.

If you Google "cool, green, wet," as my husband did, the first pop is likely to be the coastal mountains of Oregon. Eighty inches of rain annually keeps it wet. Moderating influences of the Pacific Ocean keep it cool, never cold or hot. Towering doug-fir forests keep it majestically green. And that, roughly, is how you get from lucrative careers with comfortable lifestyles in Phoenix to a sheep farm tucked into a slot valley in the Coast Range of Oregon. Well, that and just the right amount of delusional thinking.

Most people recruited to farming come with romantic delusions about the rural life. That's as true for the original trekkers of the Oregon Trail as it is for the latest back-to-the-land neophytes. Without the romance, there would be no new blood in farming. Of course, the irony is that nothing will cure those delusions faster than the act of farming. Nature is both relentless and unforgiving. After five years, the delusions are gone and so too are a good many of the once-hopeful recruits. What remains depends on your answer to that most basic of questions: what gives meaning to life? What follows is my account of our first five years—those years that test and reveal.

BEGIN BY ENDING

When you're twenty, adventure begins by throwing your bags in the back of the car. A quick glance to your partner, adjust the mirror, buckle up, and hit the gas. Life, like that open road, is in front of you. That's how we got to Phoenix.

At mid-life, it's more complicated. An adventure must be worked through in the movie theater of your mind until there is a story you'd pay to see. Once it's worked through, you're willing to sell tickets to it—promote it to everyone. Of course, having now paid for and promoted it, you'll find the tickets are to an entirely different movie. At least that's how it appeared to us, judging by the reactions of our family and friends. While we expected trumpets and Goldwyn's lions announcing our mid-life movie, or at least a majorette with sparklers on her twirling baton, that's not what we got.

The reaction began with our children, young adults vested in colleges as far from us as they could get. Our adventure was their betrayal. The house they couldn't wait to leave was now the sacred shrine of all their childhood memories. We were the curators they had entrusted to keep it safe. Instead, we sold it to the philistines on the cheap. They were not even consulted! Well, not sufficiently enough to believe their parents were serious.

And they were right. With the children gone we had begun to think of ourselves as a couple, and the last time we were a couple, we were twenty. Throw the bags in the car and hit the gas.

It continued through our extended family and friends. The same people who suffered through Phoenix summers, complaining of gridlock, dreaming of an escape, now asked, "How could you?" instead of, "Save a place for me." Of course they did. Our adventure was their loss. And their loss ultimately was our loss too.

At mid-life, adventures begin with endings. That's because we don't age by becoming set in our ways as much as we become set in our relationships. Each year adds another layer to the network of support we all depend on. By mid-life, our social fabric is so tightly bound that any tear caused by a prospective move would, if properly measured, scuttle most mid-life transitions. In our material world, it's often the value of relationships we underestimate. A simple economic formula can decide whether you take or leave the sofa or box springs, but relationships always stay. Always. Relationships exist in a certain time and place and there they remain.

As painful as this truth is, it is not a reason to stay. Change is inevitable—even for those relationships that remain in place. To not heed the call to adventure has its price too and must be measured. Regret can shrivel lives and stunt relationships.

And so we said our good-byes, comforted by the thought that although our relationships would stay behind, our memories were portable. Not only portable, they were selective as well. Going forward, we would carry the memories of everyone who loved us and all that we loved. Just as farmers hold back their best seed for next year's crop, we selected only the best memories to nourish our relationships. Memories, properly nourished, take root in our daily action, becoming part of who we are. We were not a young couple; we were mid-life with lots of memories. And, thanks to modern communication and transportation, there would be opportunities to refresh those memories. Your family and friends may not move with you but they will visit.

The final good-bye was for ourselves, or more accurately, for our "selves." Every move involves the sorting of stuff, and each micro-decision asks you to define the new you. Throw out the business suits and keep the T-shirt, you're going to a farm. But the black cocktail dress was so elegant and I only wore it once! What parts do you keep for the new you that has yet to emerge? It doesn't matter, because it's only a guess. We will always be wrong. Like the pianos discarded by the side of the Oregon Trail, the journey will decide for us.

But, if ever you come upon a farmer in a black cocktail dress, feeding her chickens—consider she may not be crazy, just transitioning.

OREGON GREEN

An hour past dawn and the temperature was already spiking above 80 degrees, dashing our hopes of making it off the desert floor before the bake set in. We traveled not by car but by caravan—each vehicle had a position, a mission, and a shortwave radio. June is an auspicious month for new beginnings, and the beads of sweat only polished our resolve. Oregon—cool, wet, and green. Part mantra, part sacred promise.

I drove point. My job was to scout the various pit stops and report on the road ahead. My companion was Bezel, a crotchety old cat who complained incessantly. Greg, my husband, piloted a land-barge of a truck piled with stuff and trailering our two steeds: Chaco and Mora. Our two girls, Caitlin and Annie, drove behind the trailer making sure its anxious occupants were not creating problems. Assisting them with navigation, air flow updates, and squirrel sightings were our dogs, Patches (the good dog) and Cisco (the bad dog).

The adventure was underway. I couldn't help but wonder at possible similarities between our caravan and the original Oregon Trail trekkers. Like Greg, they must have been full of hope for the future, but I wondered if some of the children were resistant like ours. Our oldest, Caitlin, had adopted an adversarial attitude while Annie remained disengaged. I was attempting to persuade the girls to stay open to the experience. I doubt that nineteenth-century trekkers dealt with vocal objections from their children, but I bet the tensions were there just the same.

Of course, the similarities end there. The historical trek entailed months of hardship and deprivation, delivering the family into a scenario of enforced cooperation to survive. Our girls would be back in college at the end of summer.

The shortwave buzzed, asking me to adjudicate their squabble over music selections. My children had an abundance of choices and none of them related to their survival. These were twenty-first-century girls learning to use their voice. I suggested they had the resources to solve it themselves. It's challenging to keep the bonds of family when there are so many choices.

Two days of truck-stop food, car camping, and bickering children is hard enough, but with a traveling circus of stressed-out animals, it took grueling to a new dimension. We consoled ourselves that our animals had the most to gain from the trip. These backyard ponies would have sixty acres of Oregon pasture to find their inner mustang. The dogs would have endless woods full of verdant smells to delight the canine nose. And the cat would have an actual leafy, green tree for plotting a full-on, feline ambush. Maybe this move came with some costs for us, but for our animals it would be a return to Eden. Our animals were going home to the country.

We arrived the morning of the third day. The sun was just starting to peek over the mountains, casting long blue shadows. It was chilly in the shade, but where the sun hit, there was promise of a beautiful day. We led the horses, wild-eyed and snorting, out of their dark trailer and into the abundant green of Oregon. We loosed their halters and watched them run—in sheer panic. They looped the field several times at a full gallop until they could no longer sustain the pace and slowed, frothy and sweating, to a hoppy canter. We realized we were going to have to catch these two hot-heads and walk them off to prevent them getting chilled. Catching them with sixty acres of sprinting room would take some doing.

Before we could put that plan into action we heard the agitated screams of our girls. Cisco, the bad dog, was making for the woods with a chicken in his mouth. Both girls were in hot pursuit. Patches, the good dog, had dutifully dropped her chicken when commanded but was now accompanying Cisco in his flight from justice.

And Bezel, the cat, who complained incessantly about riding in a car, was balled up under the front seat refusing to come out. It would require heavy-duty leather gloves to overcome his objections.

Thus concluded the first hour of our new life on the farm—and with it came the puncture of another delusion. We assumed our animals would be at home in nature. Instead, they were unbalanced and frightened by it. Like us, they were "at home" at home. It's familiarity that instills comfort. It would take them weeks to get their bearings, and months to become country animals. Until then, their lives were in peril. Coyotes and owls consider city cats the country equivalent of fast food. Dogs that run at livestock, including chickens, are shot on sight. And thousand-pound horses can tangle in barbed wire and shred themselves to the bone in minutes. We were naive to presume to know country ways for our animals.

Green is the color of Oregon. It reflects the botanical abundance of the state and is represented in everything from the license plates to the football team at the university. It is also an apt description of new immigrants to the state. That day we learned the truth of Oregon green—it applies to animals as much as people. And for both, it takes lots of experience to replace the green with bark.

DAY ONE

People sell homes because they want to move, usually to better digs. People sell small farms because of personal tragedies: sickness, bankruptcy, divorce—the reasons are never good. And while it would seem that selling a farm should not be much more complicated than selling a home or a business, it is. Since farms are both residences and businesses, the models for calculating valuation are thrown into flux. And without reliable valuation, banks will not lend money. The smaller the farm, the less profit, the more difficult the valuation, the harder to get financing and, ultimately, the longer the farm is on the market. Bottom line, most small farms come to new owners having languished through long periods of disrepair and neglect.

While our animals were getting acquainted with the farm, the farm began to introduce itself to us. And like any jilted lover, we were about to encounter its scorn. The electricity was on, so we had lights and water, but that was about it. No phone, TV, computer—no access to the wider world. The stove and the refrigerator were shot, so meals were largely PB&J sandwiches.

Farmhouses, especially old ones, were not intended to keep nature out, more like slow it down to a manageable level. There was a good sampling of the rodent population in various stages of decay randomly distributed through the house. Removing them only seemed to invite their living relatives to return and re-populate. We had hoped our cat would be a deterrent, but in protest to the move, he had made off to the darkest corner of the house and curled into a "do not disturb" ball. From deep within that black fur ball we caught the glimmer of half-cocked eyes and exposed fangs. Clearly we were on our own for staring down the rodent horde.

The farm itself was "free range," not by intent but by default. Chickens and sheep went where they pleased, much to the consternation of our dogs, who were now either locked up or leashed. We made a mental note that fence repair would be among our first priorities. While the sheep roamed, our horses hovered at the barn. They were in danger of dehydrating less than one hundred feet from a stream. Being desert horses, they had never heard rushing water and were terrified by it. They had to be hand watered from a trough.

After taking stock of our surroundings we began the ordeal of unpacking. By mid-afternoon the heat and sore backs had slowed our pace and must have suggested our vulnerability. A beaten down pickup hovered on the county road, then eased down our driveway. In the cab was an older man constructed of knotted cables and burlap. He had an erratic gray beard, wore a T-shirt with the sleeves torn off, and supported his faded jeans with red suspenders. After surveying our move-in, he announced in a voice accustomed to talking over chainsaws, "Looks like someone threw a hand grenade in the middle of a yard sale."

Greg's initial instinct was to suggest another place where the visitor could stick his hand grenade, but one look at the big man's goofy grin and we were disarmed. After introductions, the visitor swung out of the cab and headed for the tailgate, which he dropped. We had just been invited into his parlor.

"You're supposed to welcome new neighbors with an apple pie but I thought you might appreciate a cold beer even better."

And we did.

"Name is Rick Reahl, but everyone calls me Brick. I like to think it's because I'm solid built, but could be 'cause they think I'm thick in the noggin." He thumped gnarled knuckles against his bald dome, made a goofy face, and laughed.

Brick and his family lived at the end of the road in a little cabin built by his grandfather some hundred years ago. You knew you were at Brick's place when you came upon a dozen signs, posted at varying eye levels, warning of dire consequences for the sin of trespassing. This means you! There were not a lot of people who made it to the end of this obscure road, but those who did were properly warned.

A bad back eventually retired Brick from logging and now he did odd jobs between fishing and hunting seasons. We learned later that in Oregon there are no days between some form of fishing or hunting season. Brick's real occupation was the mayor of Hopping Frog Road. Everyone knew Brick and everyone liked Brick. In small towns, where casual slights can build into generational animosities, that was a real gift. He moved fluidly between the redneck logging clans and the eco-hippy tribes, and he did this largely by keeping confidences and engaging the positive side of each individual. And he made it seem just that simple. For us, he gave leads on people to help with farm repair and a brief orientation to the town. Then the tailgate shut and off he went to other mayoral duties. We went back to moving boxes.

That night we rolled our sleeping bags out in the living room and polished off our PB&Js. Another night of camping, but at least we had lights and running water. My reward for the long day was going to be a hot bath to ease aching muscles. When I opened the tap, it spit out a dollop of red water followed by hissing air, then silence. I just stared, then I cursed and banged the pipe, then I cried, and then I stared some more. The horses! I had left the water running in the trough, draining the holding tank. We were out of water. No water to wash. No water to drink. No water to flush.

For city dwellers, light and water are taken for granted. When you open a tap there should be water just as when you get up in the morning there should be a sun in the sky. In the country, water and lights are more a hope than an expectation. What you can "expect" in the country is that at some point in the near future you will be without one or both.

Flashlight in hand, I trudged to the barn and turned off the valve at the trough. As I would soon learn, I did not need to do this because the pump had lost its prime and would need to have the air bled out of the pipes before water would move through them again. Before that, the spring would need to refill the holding tank. With luck we'd have water by tomorrow.

Well, at least we had lights. Naturally, the lights began to falter just as this thought occurred, warning me, "this too can go away." Yes, I was making the acquaintance of Mr. Farm. He could be insistent.

As I drifted to sleep that night, a dull thought bubble percolated up from the mud of misgivings in my pre-conscious mind: *What have I done?* Just as it burst, I realized it wasn't me. *What has my malcontent husband done!* This could be a terrible mistake. And before these thoughts could rouse me from my slumber, another thought bubble dislodged from the muck: *my parents!*

I remember the phone call when we announced our plan. The long pause. My father clearing his throat. The polite inquiries into the state of our sanity. My mother's silence. My father's strained attempts at support, followed by his attempts to solicit my mother's support for this plan that neither of them could understand. My mother's continued silence cloaking her secret thought that my husband would be the ruin of me. And what could be more ruinous than the isolation and drudgery of a farm? It all culminated in my father pledging they would assist us with the move—perhaps to force my mother to a more supportive role, perhaps to signal to my husband that their daughter was not lost to them, perhaps to help us find the mental health services we so desperately needed.

That was the final, cold thought that streaked across my pre-conscious brain. My parents were arriving in forty-eight hours. We had two days to get the farm sufficiently presentable to allay their worst fears. Yeah, that's likely.

DAY TWO

Our first farm chore was to get the sheep off the lawn, which meant repairing the fences. The trick to fencing is having a secure base to draw the wire tight. Posts only augment the stability of the fence and maintain the spacing of the wires. The H-bar is the most common solution for a secure base. It consists of two heavy posts sunk three feet in the ground with a board nailed between them (forming an "H") and two loops of wire diagonally transecting the two posts. A stick is used to tighten the wires, which brace the posts against each other, creating an immovable object. The fencing is secured at the H-bar and then stretched across the row of posts until it is secured at the next H-bar, usually at a gate or bend in the fence line. Stretching the wire can be done with a tractor or a come-along. It's best to use two people, especially if one of them doesn't know what he is doing.

Greg hired Russell, a young, beefy farm kid, to help string the fencing. By midday, Russell was looking down a long row of posts and decided to pick up the pace if he wanted to keep his date with his girlfriend that night. With the vigor of young love, he drove his shovel through a four-inch irrigation pipe that wasn't supposed to be there, creating a geyser and a whole new work priority. When the irrigation goes down, the grass stops growing and the sheep start ranging beyond the unfenced pasture. This will not please your neighbors.

The irrigation repair requires digging up the pipe buried two feet in the ground, cutting the break, slipping a coupler over the repair, and gluing it. Bending large pipe, buried in the ground, sufficiently so a coupler will slip over it, requires a lot of digging and some engineering. And there's the trip to town to get the supplies. Tick tock. Grass is dying. Sheep are ranging. And the farmer? Well, this farmer is having an existential crisis. Surveying the geyser, his only response to Russell's spontaneous "oh shit" was, "What have I done with my life?" He's been farming less than twenty-four hours.

My assignment was to get the farmhouse operational. I had driven to town to procure a stove, refrigerator, and phone connection (no cell phone service in a slot valley). It's easy to buy appliances. Getting them delivered to a rural address requires a week's notice, a cooperative retailer, and a substantial delivery fee. Cooperative retailers proved to be in short supply.

First Law of Country Living: The first thing you give up by moving to the country is convenience. There's no pizza delivery down a gravel road.

The phone company was the most receptive, probably because it's a rural cooperative and has empathy for the plight of rural folk. The petite older woman at the front desk took one look at my unwashed hair, dirty jeans, and desperate face and guessed the rest.

"New in town?"

"Just moved to Hopping Frog Road."

"Not the Spences' old place? Well, it is beautiful out there."

I would learn later that this is Oregon code. Oregonians live by the rule "if you can't say anything nice . . ." "Beautiful out there" meant there were no other redeeming qualities. She moved my name to the front of the work orders if I agreed to be home by mid-afternoon. I swore a blood oath.

"What number would you like?"

I was perplexed. "What do you mean?"

"What number would you like for your phone? We find it's easier for people to remember if they pick it themselves."

I had never been presented with that option. Come to think of it, I had never actually talked to a real person when it came to any dealings with phone companies.

"Uhh, I don't know . . . given the day I'm having, anything with 666 seems to fit."

"Oh no, the headbangers got that locked up till the second coming. How 'bout 9966? That's kind of easy."

"Sign me up."

There's nothing like the isolation of country life to make you appreciate a phone. For me, it meant fewer trips to town. But for my girls it meant a return to civilization. So the lanky young lineman was greeted as a hero by the twenty-somethings in the family. When he realized he would be crawling under an ancient farmhouse full of nasty bugs, rodents, and possibly a rabid skunk or coon, he began to feel entitled to the laurel-leaf crown. When he viewed the tiny, dark spider hole he had to squirm through, his mood darkened considerably. We agreed to leave him to his task and take the dogs for a walk.

Since we were walking through the woods and away from the livestock, we decided to unleash the dogs. At last, they were free to engage the leafy abundance around them! It added to our joy to watch their excitement. That is, until they caught a scent and bolted off. We hadn't realized the sheep had already started to range beyond their pasture. We raced through the woods following the sounds of bleating and barking. The rout ended in the creek. An old ewe was in the water up to her belly and backed up against a steep bank. One dog had hold of her ear and was pulling her down while the other was lunging at her with vicious snapping jaws. When pack mentality takes over, dogs readily convert to their wolf archetype. We had to use sticks to drive them off.

Free from the dogs, the ewe just stood in the creek, earless and bleeding profusely from a number of wounds. She was in shock and would not move, so we were forced to drag her up the embankment. A full grown ewe weighs about 150 pounds, but with wet wool you could add another 50. She lay on the bank, limp, in utter resignation of her impending death. We cleaned the wounds as best we could. Annie, kneeling beside the ewe, began to sing a gentle lullaby to soothe her. The sound of the creek and Annie's soft, clear voice shifted the tempo from crisis to comfort. A small clutch of ewes lingered nearby. I drove them over, thinking the herd might inspire the ewe to get up. Sure enough, she hoisted herself up onto quaking legs. We pushed the group up to the barn. There, we were able to isolate the injured ewe in a stall. Good. Now all we needed was a vet. Well, a phone and a vet.

The hero lineman had crawled out from under the house looking exhausted and relieved. We now had a phone. I searched for the name of the farmer down the road that Brick had given me and placed a call. So, this is how we meet our neighbors—through crisis. The farmer's wife answered.

"You don't know me. We just moved onto Hopping Frog and our dogs attacked a ewe. We need a vet."

"A ewe?"

"Yes, she's bleeding pretty badly."

"A ewe costs fifty dollars. Just to get the vet to your door is a hundred. Clean the wounds. Tomorrow the ewe is either alive or dead. Either way, you're still fifty dollars ahead of a vet visit."

And there it was: the Second Law.

The Second Law: You probably can't afford experts, so improvise. If that doesn't work, call a few people and improvise again.

We took the advice. The ewe was still alive the next morning, so we released her back to the herd. She became our earless (deaf) ewe and went on to lamb for many more years.

SHEEP SHORTCOMINGS

To prevent more earless sheep, we decided to drive the herd into a temporary pen. It is said that sheep are the leading cause of divorce in rural America. Hard to believe, when looking at the bucolic scene of sheep grazing serenely in a pasture. But try moving them toward a goal. Pushing sheep is roughly akin to pushing water with a sieve.

It is also said that sheep are among the dumbest creatures on God's green earth. It is usually said by a farmer being out-maneuvered by his sheep. And if your spouse is helping you move the sheep, it becomes abundantly clear that the only thing on that field dumber than the sheep is . . . well, that's probably why shepherds have dogs as companions.

So what's so hard about moving sheep? They're prey animals, which mean they have almost 180 degrees of vision out of each eye for viewing threats. Anything moving toward them that signals a threat causes them to bolt. If they're packed in formation, they bolt in all directions, circling around until they pack up again. The sheep's strategy is, if you're in a pack, maybe the guy next to you will get eaten and you'll get away. Unless the predator is too close, in which case, you're the sacrifice. Then, better to just bolt. So the pattern is pack, bolt, pack. The trick, from the sheep's perspective, is detecting a predator from the casual passerby, and that has to do with speed and singularity of approach. Of course, any human approaching a herd looks a lot like a predator, causing the herd to bolt, circle, and regroup behind them. Now, walk in a large circle around the herd, then approach, and watch the process repeat. You can spend many hours of your life playing the circle game with sheep. They don't mind.

To move sheep, you must approach just slow enough to cause them to feel uncomfortable so they want to move, but not so fast as to trigger their flee response. Often, you will have to stop and patiently wait for the herd to calm down before proceeding. There are long moments of gazing into each other eyes, trying to read each other's intent. The instant the herd puts their heads down to eat, take a gentle step forward. You can do this for half an hour or so and then, oops, that last step was just a bit too soon, or too large, or too straight, and off they go, back to the starting line.

It is very, very hard to move a herd by yourself. It's just too easy for the herd to keep circling you and coming back to the beautiful grass you want them off. If you're moving a herd with a partner, you must be in as perfect sync with each other as the herd is with itself. You must each stop and start in harmony, constantly working that delicate threshold of uncomfortable, but not a threat. You must sense the breathing, the heartbeat, and the tension of the herd and, with the quiet grace of a Buddhist monk, in unison, apply just the right amount of tension.

That's why successfully moving sheep should be a requirement for granting marriage licenses in every state in the union.

And the reason sheep are the leading cause of divorce in rural America? It's very hard to assess who or what caused the herd to bolt, making the situation a kind of kinetic ink-blot for projecting blame. If you're already stressed and thinking of all the things you have to do and haven't done and feeling inadequate and/or unsupported, meaning you're farming, this is just gasoline to the match. To preserve your marriage it would be better to blame the sheep, except your partner is supposed to help while the sheep are just being sheep. There's nothing like sheep to reveal the shortcomings of a marriage and the people that inhabit that relationship.

After several sharp exchanges on the differences between gender and intelligence and a few more stand-downs to regain composure and retain marital status, the sheep were at last contained! Except for that one ewe in the far field, which we saw as soon as we turned back. How could we have missed it? I was already calculating who was responsible for collecting sheep from that side of the field, when I noticed something even more troubling. What was that stick figure wobbling behind it? On second look, there were two stick figures rising up out of the grass. Lambs! Oh, God, we've only been farmers for thirty-six hours. We're too young to have lambs! We both looked at each other hoping the other knew what to do.

Back to the house and the Internet to read up on sheep. Then back to the field to grab the lambs, still wet and bawling. Show momma her babies and she will follow you all the way to the barn. We put her in a stall with fresh straw to ensure the lambs were warm and nursing properly. Heads down at the spigot and tails up and wagging—sure signs all's right in the world. A few days of bonding and then back to the herd—well, after another call to the farmer's wife and some much needed assistance with weighing lambs, giving shots, docking tails, tagging, and (since they were our first) naming—Rosy and Rudy. Welcome to the world.

We could just as well have said to ourselves, "Welcome to farming." We too were lambs being initiated into a new world. Over the course of the morning we had gone from a goal and a plan to get there, to frustration, to infuriation, to surprise, to helplessness, to improvisations, and finally to the bliss of new life and a profound feeling of accomplishment. And our reward? Back to the chores. Welcome to farming.

By evening Greg had finished the repairs to the fence in one pasture. With the use of a phone, I was able to find two young men to deliver the appliances and a local, retired electrician to hook them up. Things were falling into place, and to celebrate, we had our first home-cooked meal. My parents were arriving tomorrow and I was beginning to feel ready for them. I knew they were worried for me and I wanted to dissipate as much of that anxiety as I could. I wanted them to like the farm, the way we liked it when we bought it. Up until this moment, it had seemed the chaos of the past few days made that unlikely. So, for the first night since leaving Phoenix, I headed to bed with a feeling that things might work out.

A piercing scream ended that thought. My daughter, emerging from a steamy shower, saw eyes pressed up against the window of our second story bathroom. By the time I got there, they were gone. She was too startled to give a clear description except they were "dark." The window was filled with empty blackness and beyond was dense, brooding forest. We were alone, the nearest neighbor too far to be of practical help against an intruder. I tried to calm my daughter, but peeping into a second floor window scores pretty high on the creep meter. Just as she regained composure, a shaft of bright light shot across the window and we both screamed.

Greg began calling to us from outside. Caught in the beam of his flashlight were three masked interlopers perched in the mulberry tree. We looked at them and they looked at us until the mama raccoon called her little delinquents up the tree and out of harm's way. Slowly they lumbered further up into the leafy darkness. My daughter, wrapped in her bath robe, was cooing, "How cute!" In that moment, the mood had shifted from the terror of the unknown to the wonderment of nature. We were witnessing the same phenomena, just through a different frame for viewing it—well, that and a good flashlight.

We all headed back to bed, but I began to realize, that feeling that "things will work out"—it was short lived.

THE THIRD LAW

The second law of thermodynamics states that "everything turns to chaos." Okay, that may not be exactly the second law of thermodynamics, but it is the Third Law of Farming. The first two being: nothing is convenient, and you can't afford the expert so improvise.

Nowhere does the Third Law apply more appropriately than to irrigation. There's something about moving parts and water under extreme pressure that destines irrigation systems toward chaos. Factor in small farmers on tight budgets buying used equipment from different manufacturers (that don't actually fit together) and you get the picture.

Our farm, with its different pasture configurations, uses two systems of irrigation: a long row of thirty-foot pipes, each with two-foot risers, and a rain gun that shoots a lethal spray of water one hundred feet in a circle. Both systems have to be disconnected, moved, and reassembled daily. This was Greg's first chore in the morning. On the third day, he dropped the rain gun, creating a slight bend in the rocker arm of the spray head. That's all it took to hit chaos.

We didn't know it quite yet because we were distracted by the arrival of my parents. They were easy to spot—the only Volvo driving a gravel road where Dodge Ram trucks with Cummins engines were de rigeur. My father wore breezy lemon-colored slacks with a plaid shirt and a fly-fishing cap while my mother sported the latest L.L.Bean country couture. After visiting and a tour of the farm, we parceled out chores. My mother is an avid gardener, so we headed to the nursery to begin my instruction in the cultivation of botanicals. Dad accompanied Greg to the Merc for irrigation parts.

The logging hamlet of Elsie is little more than gas station, post office, library, school, diner, and the Merc. Short for *mercantile*, the Merc was more than a country store—it was the beating heart of Elsie. The one place everyone in town would visit at least once a week and therefore guaranteed to have the latest news updates. Almost any life necessity could be found in its two thousand square feet of clutter, including: food, clothing, ammunition, chain saw oil, pinot noir wine, fishing licenses, lumber, oil lamps, rubber boots, hardware, cooking utensils, hand-made furniture, chicken feed, and, yes, irrigation supplies. Stocking a country store is an art that requires intimate knowledge of the needs and desires of the community it serves. That rare gift for intuitive assessment resided with a burly, mohawk-sporting, square-jawed ex-Marine named Nate.

On the day I met Nate, he was crouched behind the counter. He had rigged a cardboard grizzly (advertising beer or chewing tobacco, I forget which) to pop up whenever he pulled on a string. This would send small children screaming down the aisle until they felt safely out of bear reach. Then, they would begin their creep back to the mysterious paper bear. Nate winked at me to keep his secret as he tugged on the line again to another wave of exhilarated screams. All commercial transactions were on hold until the last scream had been milked from the last waif.

One overworked mother tried to clue me in. "We're waiting for our children to catch on. Some never do." Her head nodded to the large man with a mohawk crouching behind the counter, still chuckling after the children had caught the bear and exposed its string.

So, when the irrigation broke, Greg's first stop was the Merc. Nate was a seasoned problem solver, but he took one look at the rocker arm and referred Greg to a professional. Greg was too naive to know it yet, but any problem beyond Nate's expertise was clearly in the realm of the Third Law. The good news—there actually was a store that specialized in irrigation. The bad news—you had to drive to the next town to find it. Remember the First Law—nothing is convenient.

In the city, retail is a fairly simple matter of selecting the product you want, paying for it, and leaving. In and out, as quickly as possible. In the country, retail is a much richer experience. Unfortunately, Greg was still expecting the efficiency of the city when he entered Doug's irrigation shop.

Doug, possessing a big frame with powerful arms accustomed to bending pipe, stood behind the counter with glasses perched on his deeply lined forehead. He was a serious man in a profession that required the deep analytical skills of an engineer.

Two customers preceded Greg. The first was giving Doug a fishing update. This led to recollections of great fishing events in the past, which led to discussion of the best river access points, which led to land use laws and excesses of government regulations, which led to critiques on the conduct of politics (both local and national), which led to broad observations on the status and direction of American culture, which brought the discussion back to the problem at hand: "We need a part, do you have it?" Each topic broached much opining from all parties present. The second customer repeated the format but selected elk hunting as his entry point. Greg kept hoping an employee would magically appear to open another register.

When it was finally Greg's turn, he dropped the rocker arm on the counter and asked boorishly, "Can you fix this?" Doug looked at him to size up what kind of a rube he was dealing with, then looked at the arm and asked, "Where do you live?" After the long wait for service, this question did not reassure my husband that he was transacting with a competent professional. The relationship was off to a rocky start.

The little dent in the rocker arm was a big problem—largely because the manufacturer had gone belly-up over twenty years before. So, the first question was, "Could the bend be straightened or could an obsolete part be found in some farmer's field?"

The customers weighed in. A farmer in Linn County used that type of rain gun. He might have a part. Another thought a repair might work. Doug was doubtful. The alignment had to be perfect. He would take it under advisement, make a few calls, and get back to us in three days.

Three days without irrigation—without graze for our livestock! Now Greg understood, we were officially in chaos. Fortunately, Doug knew what that meant for his customers. He suggested we supplement with hay from the farmer down the road. He even placed a call to the farmer to confirm.

Doug did not know us, but he knew our farm in all of its complexity. And he knew the farms around us. Those long fishing stories are how people stay connected. Farming is a complex business with complex problems, and having access to a wealth of resources is key. Staying connected is key. People are drawn to farming by the allure of independence. "Grow your own food, no one to tell you what to do." People survive farming by mastering the art of interdependence.

Fourth Law: Farming involves nurturing ever greater levels of interdependence.

And so Greg left Doug's shop not with a part but with a connection to a neighbor. Now, if he had a map, he could have spent less time searching for that neighbor.

Will Ridout, a craggy man, tall in stature, severe in nature, and short on words, met Greg and my dad at his barn. He said nothing of my father dressed in his lemon trousers, just directed them to pull the truck around so they could start loading hay bales into it. Then, they drove to our barn, where the craggy man lobbed fifty-pound bales from the bed of the truck to our second story loft while Greg and my father stacked them. At fifty-five, Greg was the youngest man on the buck line, something of a testament to the aging of farmers in America (farmers over age sixty-five outnumber farmers under thirty-five by a ratio of seven to one; 32 percent of beginning farmers are fifty-five or older).

When the unloading was done, Will thanked us for buying his hay with a big smile. He may have been short on words, but he was big of heart and generous with his farm wisdom. We didn't know it yet, but we had just made a valuable contact. I think Doug anticipated that outcome. Like most people in the country, Doug and Will knew the Fourth Law of farming was the answer to the Third Law. And it's why the correct response to "Do you have a part?" depends on where you live.

SO THIS IS FARM LIFE

There's nothing like farm chores to produce tired muscles and deep sleep. On this night, a little too deep. While we slept, Mrs. Coon took her young ones on a savage killing spree through our hen house. Matching the carnage strewn across the chicken yard that morning with the cute bundles of fur in the mulberry tree of the previous night was too incongruous to believe. Maybe it was a bobcat or a weasel? Maybe, but no bobcat or weasel had been seen, and the coons were a known quantity. Enough circumstantial evidence to convict them in any court of rural experience.

The surviving chickens were lined up and cackling for their breakfast. When it comes to brains, chickens have neural tubes, which is why there would be a repeat performance tonight if we didn't fix the holes in the wire fence. Survival instinct is the first thing to go in the evolutionary process toward domestication. Securing the chicken coop had just risen to the top of the chore list.

Before we could complete our assessment of the repairs, a fluffy white line of grass mulchers ambled across our lawn, nipping every flower in sight. The hen house repair just got bumped down. We heaved a collective sigh as we moved to head them off at the pass. This time, I pledged, I was not going to lose the contest of wits with God's dumbest creature. I was not going to lose my cool. I lost that pledge by the time I'd driven the sheep to the bottom of the hill.

At the pasture, we could see the gate was off its hinges. During the night a ewe must have stuck her bony head through the slats of the gate for that greener grass on the other side. When she drew back, it snagged, causing the ewe to jerk her head and lift the gate up and off its hinges. That or she just stuck her head into the slats and lifted it off on purpose—a theory of sheep chicanery espoused by some in my household. Country gates are hung on simple pins with cylinders over the pin to allow the gate to pivot. Pointing one pin down before re-hanging the gate solved that problem. Back to the hen house.

We were midway through the coop repairs when our daughter informed us the water was out. Again. An inspection of the spigot at the trough eliminated over-watering as a source of the problem. The pump was operational, but the holding tank was empty. Either the tank had sprung a leak or the problem was with our spring. No mud around the tank left us with the most troubling possibility—the spring itself.

The house water came from a spring flowing out of the side of the hill. The water from the spring was collected in a spring box and then piped to the holding tank. A cave-in had buried the spring box in mud. After a day's labor, Greg had dug out the box. Next he cleared the pipes, first by blowing and then, when that failed, by sucking the mud out.

Fifth Law: Farming is a dirty business; at some point you will eat dirt and probably a lot of it.

The water began flowing back into the tank, but once again it would be a night without running water—without baths. It was too late to finish the hen house so we resorted to a temporary fix. That evening, Doug phoned to say the repairs on the rain gun would take a few more days. The Third Law of Farming is as predictable as gravity.

The next morning arrived with the water running. I was measuring coffee and gazing out the window when a fluffy white ball began mulching its way across the front lawn. Soon it was joined by others until they formed a line that could have been the envy of a marching band. I had one solitary thought—*loose the dogs*.

As I continued to stare, a sinking feeling took hold. No matter what we did or how hard we worked, nor how smart, talented, and committed we were, nature would win. It was a dark thought of total capitulation. I decided I needed my coffee and a moment more to regroup. The sheep could wait.

Greg came in and asked, "What's that?" I couldn't bring myself to answer. We sat together sipping coffee and taking in the scene. Finally, Greg put his cup down and said, "Better get at it." We rose together in stoic syncopation as old as farming itself. American Gothic.

But part of me resisted. Part of me asked, *is this really what I want?* I feared we were transforming into the characters in the painting, pinched by a life of hardness, deprivation, narrowness of expression, and menial labor. We were crossing the line from stoic to gothic. That thought now took possession of me.

It was the last day of my parent's visit and I promised my father a ride in the back country. The horses, still unaccustomed to the trees, were snorting and side-stepping and trying to turn back most of the way up the old logging road. Frankly, I didn't blame them. The forest was dark and foreboding, overgrown with green ferns. Moss hanging from the limbs gave the trees a sinister appearance, as if nature herself might drop her green shroud and vanish us into musty oblivion. But when we cleared the summit, we encountered blue sky and a majestic view over the valley.

"Gosh, would you look at that. It's a bit of alright."

My father came of age in the 1940s and never abandoned the vernacular of his youth. I agreed and sat quietly taking it in, but in the back of my mind was that lingering question: "Had we just made the biggest mistake of our lives?" We knew it was going to be hard but not this hard, not this relentless. And we didn't seem to be very good at it, as attested by the continued free ranging of our woolly horde. And the expense! All these repairs had us hemorrhaging money. This was starting to create tension between Greg and myself. No, this was not what we signed up for, definitely not what I had signed up for, and now it was going to take us years to sell the place . . . and we'd go bankrupt . . . and divorce . . . and . . . we'd still own this damn farm! Like a lead weight dragging us to the bottom.

I was thinking this when my father asked a probing question: "Should we turn back?" He meant should we ride back to the barn. I heard, "Should you sell the farm and return home?" I erupted in a monsoon of anxiety and tears.

My father is a gentle soul and, of course, he was made uncomfortable witnessing my pain. Wanting to help, but not quite sure what I needed, he struggled, "I, umm, wasn't that keen on your coming here, but now that I see it, well . . . I understand what drew you to this farm. It's a very special place."

"Un-huh."

He could have been comparing Mickey Mouse to Mighty Mouse—at that moment, just hearing his voice helped calm my panic. But what he was saying seemed to be helping too.

"I think, maybe ask yourself what would you be doing if you didn't have the farm? In Phoenix, it seemed like everybody was kind of going their separate ways. Maybe give yourself credit for at least trying something different."

Okay, that was a good point.

"It's a lot of work, though. I don't know. Maybe give it a little more time. Transitions are always hardest at the beginning. Maybe take some classes in farming. Isn't there an ag school at OSU? Read some books. Talk to some people. Get a little more education. Gosh, sweetie, I'm not sure what to say."

He went on about how he and mom loved us and supported us and all that blah blah stuff. My father is a sensitive, gentle man and I've always known I was lucky to have him, even if he could go on a bit. He was right, though. Transitions are hardest at the beginning. And at least we were trying something different.

These thoughts relaxed me just a bit. I noticed the light coming through the trees, filtering down to the forest floor. The sharp, sweet scent of fir and cedar mingled with rich, earthy tones of molding leaves and bark. The sound of trees, whispering their secrets, was occasionally punctuated by the trill of a squirrel or the caw of a raven. Life and color were abundant around me, and I hadn't quite noticed it until this moment. This place held such beauty, such perfect peacefulness, and I didn't have to do a thing to it. Not a thing.

And, in that moment, I decided I could be incomplete. I could never do enough, my husband could never do enough, we could never do enough. Instead, I would just do what I could, and Greg would do what he could, and let the rest be incomplete.

If that wasn't enough, I'd know soon. I didn't have to anticipate it. I didn't have to worry about it. Each day I'd do what I could, learn what I could, improve what I could, and let the rest be. If the coons ate the chickens then that's what it would be until I could change it. And I'd change it when I could. In the end, it would be okay. It already was okay. It was okay before I got there. More than okay, it was perfectly incomplete.

I came off the hill with a blessing. More than one, if I'm counting. Lucky to have my father, lucky to have my family, lucky to have this farm. Then, as if to signify that very notion, we came upon an owl in the hollow of a tree. Big yellow eyes with tufted ears, just looking at us while we looked back. I had never had a personal encounter with an owl. It was a long moment before we nudged the horses down the trail.

"Gee, that was something! Did I ever tell you my nickname in school was 'owl' . . ."

About a hundred times, but I didn't care. I was so happy to be riding with my father, on this trail, on this evening, on this farm. So much that is unfinished and yet so complete in this moment.

My parents left the next morning. My father, sounding more sincere in his encouragement. My mother, sounding less severe in her reservations. And that's as good as it gets in my family—and probably as good as it should be, given the giant leap we had just taken with our lives. It was nice to have them feeling less anxious, but of course, it was never really about their anxiety, it was about mine. And I felt I had crossed a threshold. I was learning to live with incompletes. I still liked things wrapped up in a neat box, but I was getting better at accepting when they weren't. Of course, Mr. Farm had more to say about that.

A MONTH AND MORE

By the end of the month, the fencing repairs were done. The sheep were in their pasture and the chickens were in their yard. All was right with the world. Except for the one renegade hen that suddenly emerged, waddling across our lawn and trailing sixteen chicks with heads bobbing and peeping in syncopation. Besides being cute, sixteen chicks equal sixteen chickens—pretty good return on no investment.

I decided to leave the renegade mom to her reckless ways. The next day she had a dozen chicks, and the third day she was down to four. Whoever said, "Don't count your chickens until they hatch" missed the timeline. Chicks are the bon-bons of the predator world. I tried to intervene but caught only one of the fuzzy little speedballs. Next day there were none, so I was glad I saved one. My daughter took over the task of hand-raising it. As for the renegade, living beyond the fence meant she was subject to the laws of the wilderness.

When I found a pool of blood in the barn, I assumed she had met her end. But it was a lot of blood, and on closer examination it was trailing right to my mare, Mora. A gash in her pastern, just above the hoof, had flayed open her flesh and was spurting blood. Arterial bleeding is always my cue to call the vet.

The vet took one look at the wound and began lecturing me on what I needed to know to avoid wasting his time and my money. He directed me to watch as he tied-off the artery that was pulsing blood and then demonstrated the looping stitch he used to close the wound. There was a brief quiz on distinguishing white nerve from white tendon and then some more pointers on where to administer shots. I was admonished to practice by sticking a hypodermic needle into an orange until it was one fluid motion. Since he was there, the dogs got their shots. A final warning to practice what he preached and then he was gone. Someone really needs to write a tutorial on country etiquette. "When to call a vet" needs a whole chapter.

By mid-summer our skills had improved, and the pace of chores had slackened sufficiently to allow us to take stock of the farm as a totality. Initially we had bought a piece of real estate, but we now realized a farm is an organic entity—a living organism with its own unique characteristics determined by its geography. The farm, and by consequence the farmer, are rooted to a place. Farmer and farm, each shapes the other but always within the constraints imposed by that particular place.

Our farm consists of sixty acres, half in the pasture and half in the woods, set in a small valley with a stream running through it. Perched on a small rise at the center is the farmhouse, a rambling structure built by homesteaders over a hundred years ago and added on as more kids arrived. To one side of the house is a garage and workshop. To the other side, separated by a narrow creek with a footbridge, is a large barn and dilapidated tractor shed. Behind the house, an orchard of apples, cherries, pears, and plums runs down to the stream. Beyond the garage are a chicken yard, a greenhouse, and an acre of garden. There are three pastures on this side of the stream and a fourth on the other side. Beyond the pastures rise steep hills etched with the permanently green silhouettes of fir, cedar, and hemlock. In the valley, leafy alder and maple shroud the stream keeping the water crystal clear and cool. It is as beautiful as that. Which brings me to the Sixth Law.

Sixth Law: The prettier the farm, the worse the farming.

Good farmland is found in river bottoms—flat, plain, and humid. Farms in the mountains may be pretty but the soil is poor—and so it is with our farm. It is only suitable for growing grass, but the grass grows all summer long. Cool breezes coming off the mountains keep our pastures temperate. Coastal mist collects on the leaves and waters the grass. Grass likes cool, moist weather. And sheep like grass.

For novice farmers like us, sheep are ideal. They have the lowest overhead and are the easiest "keepers" of any livestock in the US today. Since they are the only commercial meat source not raised by major agribusiness in confined animal feeding operations (CAFOs), the designation of "grass fed" and "organic" is essentially redundant. All sheep are grass fed. It also means they don't require the graining that other livestock need for weight gain. Thus their production cost is not subject to rising corn and soybean prices. Since they are raised in pastures, not CAFOs, they are not subject to diseases of confinement and don't require dosing with antibiotics. Best of all, they are among the most efficient of all grazing farm stock. They are ruminants, meaning they have four sections to their stomach, which allows them to get maximum efficiency out of the grass they eat. Further, they have a wider palate (less discerning) than cattle, allowing them to eat a broader range of pasture grasses and thrive. The same acre required to feed a single cow will support six sheep. There is a risk of over-grazing since sheep nip grass at the base, but that is handled by proper rotation of pastures.

The best part of sheep is the poop. If you just nodded in agreement, you're probably a farmer. If you're not a farmer, let me explain. The most important resource on any farm is the top soil, and the best replenishment for the top soil is composted manure. Ruminants, by running grass clippings through four chemical vats, have done half the composting for you.

This living farm is a beautiful system of complex balances. On any summer day, there are all those blades of grass unfolding their photobiotic panels to the sun where they trap the solar energy in their leaves. Then comes our woolly meadow-munchers, nipping the grass and converting that complex cellulose into muscle and waste. By harvesting sheep, we capture all that energy of a summer meadow and convert it to savory nourishment for our bodies (especially when eaten with mint jelly). By collecting the waste and turning it to compost, we replenish the soil and grow more grass.

Farmers are connoisseurs of compost. Nothing brings pleasure to a farmer quite like the smell and feel of good compost.

In compost, the farmer sees the connection between poop and prosperity—not *bling* prosperity but *life-renewing* prosperity. And sheep have really good poop. Even when it's not composted, it mixes more easily into the soil than cow or horse poop. And good poop equals good soil, equals good farm, equals good earth. That's a little simplified, but not much.

Seventh Law: Poop equals prosperity. Don't hate the poop.

HARVEST

Mid-summer is the time for spreading the poop and putting up firewood. Both of these chores anticipate outcomes months away, so they don't require urgent execution. Cutting windfall into firewood is arduous work but at least it's one chore done on the farmer's schedule. Come harvest time, the luxury of scheduling chores will seem positively indulgent.

By early September, the light broadcasts long, deep shadows in the northwest, signaling the early approach of fall. This is the time of harvest, and the pace quickens to frenetic. Most produce has a freshness window of a few weeks, and some as little as a few days. Squash, tomatoes, beans, cucumbers, and berries come in abundance and spoil just as quickly.

The Himalayan blackberry is among the most noxious weeds to invade the northwest. It is free-ranging barbed wire, with tentacles long enough and heavy enough to pull down a house. Where a thicket sets up, it effectively locks out all other plant growth and closes passages for animals. Harvesting the berries is the only payback available from this felon. Good thing the revenge is sweet. Blackberry pies, cobbler, and jam will consume the first week of the harvest season, followed by canning tomatoes and pickles, and freezing peas, beans, and corn. By mid-October harvest has moved on to apples, with pies, sauces, and cider, and then on to grapes. It's a ten-week sprint to put up the produce before it goes bad.

The harvest itself is a spirited competition. Just as the farmer is awaiting the moment of perfection when the plant has bestowed the maximum sugar and flavor to its fruit, so too is a horde of insects, birds, and varmints. First come, first served. Berries and grapes in particular are harvested in direct contention with bees and wasps.

On one particular day of berry picking I got that tickling sensation of something crawling up my pant leg. I found a suspicious bump in my jeans and gave it a crushing squeeze. Instead of dying, it stung me. In retaliation, I really squeezed it, and it really stung me. I stripped off boots and pants in the middle of the garden. A yellow wasp was letting me have it a third time, then spraying me with pheromones so his buddies could help with the prosecution. I sprinted to the house, yelling all the way.

Greg found me propped on the couch, massaging three large welts with an ice cube. I'd been stung by bees, but this was an intense, hot nerve pain like no other. Greg applied a baking soda poultice. It didn't work. He moved on to every granny and great-granny recipe we could think of: mustard packs, calamine, Benadryl, aspirin, even mud, but nothing worked. It was three days of intense pain followed by two weeks of annoying welts, and the only remedy seemed to be "suck it up." Next year the pant legs were tied off. My doctor suggested I carry an EpiPen. She thought I might be developing an allergy to the venom.

September saw the kids off to college, leaving a space in the house that only became more empty when Greg left for work. With mounting expenses, Greg had picked up a teaching contract at the local community college. I persevered with farm chores. Clouds moved in, hugging the mountains in shrouds of mist. At night, agitated spiders hung fanciful webs from fence posts and trees that, in the morning, became glistening dream catchers rimmed with dew. In the woods, the leaves were beginning to catch the bright golds and muted rusts of autumn.

On one such morning I found the remains of the renegade hen in the chicken yard. I disposed of the carcass in the trash. Greg passed by it that evening on his way in from the garage and it flipped a switch for him. He wasn't losing any more chickens. That night he patrolled the trees in the chicken yard hourly until his flashlight found the intruder. With a single shot, the coon dropped from the tree, hitting the ground with a heavy, wet thud. Still, it struggled to its feet, hissing and snarling. A second shot ended all that.

The taking of life is attended by sadness. The coon was just doing what coons do. His only crime was that he had grown too dependent on our chickens. This friction between the wild and the rural is always there—a delicate commerce where each gives a little to the other. When the wild takes too much, the rural reacts. So we asserted our desire to raise chickens without coon interference, but it was not a victory, just a momentary resetting of balances in the eternal competition for food and resources.

SIX MONTHS INTO IT

Brick had taken to stopping by every few days to check on our welfare. Often he brought the newspaper and stayed for coffee. When he saw the dead coon, he became animated. There was the customary hunting inquiry about caliber of gun, difficulty of the shot, etc. And there were stories of coon carnage to justify the killing. But something else was transpiring. Until this moment, Brick had considered us visitors. At the Merc, there was even a small wager on how long we'd last. With the spilling of blood, we had passed an unspoken initiation. It signaled an end to our romantic vision of country life and recognition of the harsh realities of the country. Death is part of farming, and there is no supermarket cellophane between you and your meat.

As if to make the very point, the butcher called saying he would be in our area that week. We were to cull the lambs and leave them penned in the barn the night before his arrival. On the appointed day, I went to the barn for the morning feeding with a heavy heart. I had invested an enormous amount of time and energy to ensure the well-being of my little flock, and all of it culminated in this day of harvest. I was filled with internal conflict. As I reached the barn, I found a pile of bloody hides. The lambs were already dressed out and hanging as slabs of meat in the van. The butcher, wearing a blood-spattered apron, was cleaning up. I wasn't prepared for this spectacle but there it was—farming. And with it came the challenge to my commitment. I drew a heavy breath, pushed down my qualms, and proceeded to feed the ewes.

As I flung a flake of hay, a hidden lamb appeared from the midst of the flock. Suddenly my conflict came back in full force.

My first thought was to grant the lamb a reprieve. But there are no reprieves on the farm. Either the lamb is destined to be a breeding ewe or slated for the table. If I was to be a farmer, then this was farming. I called to the butcher outside fearing he might leave. I rushed to catch him. Just as I got to the door, he opened it and with rifle cocked shot the lamb. The lamb crumpled and a part of me sank with it. The butcher turned back to his knives. He had more stops to make and time was wasting. I went for coffee and didn't return until I heard the van leave the driveway. I wasn't that "country" yet.

With the lambs culled, it all begins again. We made an appointment for a conjugal visit from a ram, but first we had to get the girls presentable with a pedicure. Hoof rot is a common problem for sheep in the Northwest. Mud and manure get impacted, setting up bacteria that, literally, rot off the hoof. To prevent it, hooves have to be trimmed, and to trim a hoof, you have to throw a ewe to the ground. This is done by moving the sheep's jaw forward, drawing if off balance, and then whipping its head around. The sheep's body will follow its head. It's a form of sheep judo that must be done quickly and forcefully. The slightest hesitation will cause the sheep to resist, and then it's a fight. Sheep have bony heads and, when agitated, they can bust through thick wooden stalls as easily as shot fired from a cannon. They have the same effect on knee caps.

After a few lessons from Brick we were trimming hooves, or at least starting to climb the learning curve. Greg flung the sheep and held them while I trimmed the hooves and wormed them. Trimming is no picnic either. Quick a hoof with your razor sharp clippers and you're likely to get a hoof in the face or your own finger sliced. At day's end, Greg could barely straighten up, and I had several bandaged fingers.

With that onerous chore done, our girls were ready to receive the ram, appropriately named Romeo. Romeo tentatively entered the paddock, surveying the flock. He put his head in the air and took several big whiffs and, I swear, I saw him smile. I opened the gate to the pasture and the girls all trotted past like teenagers headed to the prom. And so it goes.

The days grew shorter and grayer. Heavy rains fell all night leaving the pastures soggy with standing water. The stream became a raging torrent that swept away the footbridge and cut off access to the far pasture. The rains brought down the leaves, mashing them into the brown mud, leaving only the gray skeleton of the trees jutting into the gray ceiling of a sky that seemed no more than a foot above our head. In this dreary time, nature gives us one of her greatest spectacles.

On a November morning I followed the herd to check on the pasture. Partway there I heard an urgent thrashing in the creek. Chinook salmon were churning their way up the fast waters to their spawning grounds. Our tiny creek, ankle deep in the summer, now challenged these behemoths with water heavy enough to sweep away a horse. Logs the size of Hondas came crashing by, yet the salmon powered upstream. The epic quest of the salmon never fails to inspire. It is one of the greatest testaments to resiliency in nature, and it comes on the threshold of winter, when the message is needed most.

In two weeks, it's over. Having fulfilled their reproductive mission, the carcasses of salmon wash up on the riverbanks to decay into bug fodder. Those same bugs will feed the salmon fry hatching in about thirty days. Of course, not all dead salmon are consumed by insects. At least two were eaten by our dogs, leading to salmon poisoning. Left untreated, they would have joined the salmon in short order. Once again, Brick came to the rescue, correctly identifying the problem and steering us to the feed store that stocks the Teramyacin needed to treat it. The good news is, once a dog has been poisoned by salmon, it never goes back. Sort of like that bad tequila experience in college.

December began with a hard frost and ended with snow. Our children came back just as the snowflakes began to fall. And with each flake, our farming mistakes were hidden deeper under a soft blanket of temporary denial. We needed that psychological distance to relax and enjoy the season.

One of the benefits of living in a Christmas tree forest is that trees are easy to come by—a short trip out the back door. One of the drawbacks is that trees from the forest are not anything like the pampered and pruned trees from Christmas tree farms. So, when Greg returned from the woods with a scrawny, light-starved sapling, it was greeted with resounding disappointment. For the girls, it symbolized everything they had given up. They had not come back to their home but to a foreign place of someone else's choosing. And a dark, drafty, isolated choice at that. For me, it was another dashed expectation. This was more Soviet work camp than my ideal of a New England farm accidentally dropped in Oregon. This anemic twig captured all that emotion in one visual metaphor. The opulent array of lights and ornaments from Christmas past would have to be pared down to accommodate its scrawny branches. So much for psychological distance.

But, to paraphrase Donald Rumsfeld, "you go to Christmas with the tree you have." The girls and I set about decorating by minimizing. I had always thought of Christmas as a celebration of abundance, but this tree forced a different frame upon us—less is enough.

By the time we finished dressing the tree, our mood was transformed. There was beauty in the simplicity and honesty of this little tree. Perhaps even more befitting the birth we were commemorating.

Of course, just as the eggnog was poured and we all gathered to appreciate the tree, the lights went out. This time I was prepared. Candles and oil lanterns were lit. The wood stove was stoked. We huddled together playing board games, laughing over failed strategies and bad turns of the dice, deep into the winter night. Yes, there was something to be said for simple—and for the bonds of family and for the dark of December that brings it all into high relief.

PART TWO

GO BIG OR GO HOME?

The New Year brings resolutions. Our tradition had been to take stock of the past year and set goals for the new year. So, after Christmas we tallied up our accounts for income and expenses. The numbers were even more grim than we expected. The sale of the lambs covered about 10 percent of our operating expenses. We were running a 90 percent deficit. We reviewed strategies for both reducing expenses and increasing income. Greg had taken work off the farm, but even that was not enough. We could reduce our feed expense by putting one of our pastures into hay production but that would require a tractor—another expense. The classic farm paradox: to get out of debt you have to go deeper into debt and then hope that nature cooperates.

At the end of six months, I had an answer to my question of doing what I can and living with incomplete. It isn't enough. In particular, I needed a better revenue stream. Was that even possible on our small acreage?

"Go big or go home" was coined by Earl Butz, secretary of agriculture during the Reagan administration. He meant me. Roughly 90 percent of America's agricultural output is generated by 12 percent of America's farmers. And soon it will be 10 percent. That's a testament to the efficiency of America's industrial-style farming. Never have so few fed so many, creating a base of wealth that spreads throughout our society. Never in history has a group of people paid so little for their food, leaving more funds available for discretionary spending. Farming is economy of scale. The bigger the farm, the more it can support the expensive equipment, which reduces the cost of labor, which in turn reduces the cost of the product. The cheaper the product, the thinner the profit margins, the more competitive the market. Farming is a business, and like all businesses, it responds to the market's demand for efficiency with competitive pricing.

Our little farm could never rise to that level of efficiency. It began as a farm 120 years ago to supply meat and milk to loggers when it was too expensive to haul those supplies over the mountains. Today the loggers get all the meat and milk they want from Walmart. The problems of transporting perishable agricultural goods have now been solved at the global level, making the world flat even—or rather especially—for farmers. These mountain soils could never compete with prime, flat, Iowa bottomland.

So is Mr. Butz right? Is it time for us to go home? Sell our farm into lumber production for pennies on the dollar? The words of my father kept reverberating in my mind, "Educate yourself." What did the other 88 percent percent of farmers do to make ends meet? I didn't know but I was going to find out. *I may be incomplete*, I thought, *and yes, certainly naive, but I refuse to stay ignorant. Ignorance is a choice.*

Besides insufficient revenue, there was another issue lying in the shadows. Greg had taken a job in town. In doing this, he joined the other 80 percent of American farmers (probably

100 percent percent of small farmers) who work off the farm. He was fortunate that his credentials allowed him to secure a teaching job with a decent wage and work hours that left his summers free to farm. Most farmers take semi-skilled jobs, which pay less and require longer hours, adding to the financial strain on farm families. Greg took the job because there was no other choice if we were to survive financially. It was as simple as that, and I recall it taking less than fifteen minutes of discussion to decide.

And that's how we fell into the money trap again. Not that money is a bad thing when kept in perspective. The invention of money allows us to apply a metric to issues of time, labor, and materials, which then allows us to add, weigh, and decide a course of action to solve those issues. Unfortunately, money also moves the focus off of relationships and onto material transactions. In making this decision, we never once asked ourselves what the cost of working off the farm would be to our relationship. Odd, since our whole reason for moving to the farm was to "get back to the land, get back to each other." But, that's what burgeoning debt will do to the human psyche. Without even thinking about it, we had foreclosed on a relationship option and separated a little bit from each other.

There was one more unintended consequence. Since Greg was now working in town, I was no longer the farmer's wife; I was now the farming wife. Greg was the farmer's husband. Nothing changed in my work duties, just a subtle shift of weighted responsibility. And that too would send ripples through the relationship. I was making daily decisions that previously would have had shared input. I grew to resent the shouldering of that burden alone, just as Greg grew to resent not being consulted. In other words, we wanted the same thing and were upset the other wasn't providing it. That little dab of insanity sauce was caused by focusing on the money problem instead of the relationship. When we could have been supports for each other, we were just individuals doing our level best to dig our way out of the money hole.

It all made for some sleepless nights. Maybe the other 88 percent percent of farmers (the non-industrial farmers) were having the same farm anxieties, the same relationship strains, the same sleepless nights. I tried counting sheep.

COUNTING SHEEP TO SLEEP

It doesn't work for sheep farmers. Picturing fluffy white balls jumping over fences just caused me to try to estimate the weight of each lamb, the amount of forage in the field, and the price at market. And what are they doing jumping over the fence??? That's a repair. Whose field are they in now? I'll have to get them rounded up—but they're not cooperating, they're running away, scattering. The more I run the more they scatter.

I'm not an insomniac by habit. I actually hate to be awake at night, thinking about things. I prefer to dream my anxieties into other kinds of stories. Sometimes I even dream solutions.

That first year, I sold lambs to a broker as "feeder lambs" for ninety cents per pound. In this system, the broker keeps the lambs, gambling on a better price as the market fluctuates. Of course he's also fattening the lambs, hence the term "feeder," so the better price is almost assured. And because the broker has a continuous supply of lambs, the large supermarkets and restaurants prefer to deal with the broker. The benefit to the farmer is that you get one check—one small check.

But this is the age of the Internet, and the small farmer does not have to be at the mercy of the broker. Well, not completely. As a farmer, I don't sell cuts of meat. I sell whole animals, which is called "hanging weight." I can advertise on craigslist and sell lambs for three dollars per pound, hanging weight. But between my lamb and your lamb chop there is a butcher. And there is the rub.

Federal law requires that all cuts of meat sold to the public be inspected by the USDA, which means a federal inspector must be on the premises at all times during the processing of the meat. This law stems from the terrible abuses that occurred in meat processing plants in the early twentieth century. (Remember Upton Sinclair?) Since USDA-approved meat-processing plants are expensive to build and operate, there are not many of them, and they cater to the big producers. By contrast, there are lots of small independent butchers who process game for hunters and livestock for the farmer's own table, but they don't sell single cuts of meat to the public at large. So, when I sell lambs at hanging weight, in essence I'm selling a customer a whole lamb and giving a referral to my independent butcher who will then cut the meat to the customer's specifications.

It's a convoluted system in need of reform, but it is legal and safe. For the buyer, it's cheaper than retail but you must buy in quantity. For the farmer it's three times the profit compared to selling to the broker but only one lamb at a time. Just like in my dream, one lamb jumping over the fence until the whole flock is sold.

Or I could send my lambs to a USDA-certified processing plant and sell by the cut for an average of twelve dollars per pound. I picture lamb chops jumping over the fence. If I'm selling cuts of meat, what do I do with the rest? In my dream a snow storm blows over the pasture dropping everything into ice. A commercial freezer! Sell over the Internet, use a USDA processor, and buy a freezer to store the pre-packaged lamb.

The next morning I'm placing calls. The nearest USDA processing plant is two hours away, so I would have to transport. My lambs have never left their fields—except to sample from my neighbor's. The trip would stress them, tainting the meat with stress hormones. I raise happy lambs and I want my customers to taste that quality. I can't afford to have all my hard work destroyed by a two-hour trip. Worse, the USDA plant is booked six months out. That's six more months of hay and six months closer to being mutton.

This is where the farmer turns to her most trusted tool—the calculator. After running the numbers, my best option is an amalgam. The USDA butcher is out—too far and too time constrained. I'll sell as many lambs as I can as hanging weight and the rest will go to the broker again. Not much of a solution, more like half of a solution.

That night I felt my anxiety ramping up again. I woke to the snoring of the dogs and the farmer lying next to me—wheezing metronomes of domestic tranquility. A stark contrast to the winter storm tracking through my mind. There had to be another solution. I pulled the covers up and snuggled in against the farmer. If only I could dream it.

TURKEY DREAMING

The next morning, I was sipping coffee and reading the ag news. Price of natural gas is up. Hmmm, that means the price of nitrogen fertilizer is up, which means the price of corn is up, which means it just got more expensive to feed my chickens. Well, poop!

There was another article about a turkey farmer in remote eastern Oregon. Seems he had just inherited the farm from his father, which was unfortunate for two reasons. The farm was going bust because his father had refused to update to the new breed that tripled weight gain. Seems the weight gain so unbalanced the bird that it was unable to fulfill its procreational obligations. That meant farmers would have to artificially inseminate their flocks. The old farmer judged that to be unnatural and just plain wrong. Sometimes you have to take a stand, also called bankruptcy, in business. So the son inherited not so much a farm as a zoo for a vanishing sub-species of domestic fowl. Worse, and this is the second reason, the son hated turkeys, improved breeds or otherwise.

This might make a good human interest story for the average urban newspaper, but this was an ag paper. Turns out the new breed produced triple the meat at one tenth the flavor. While this trade-off was acceptable to groceries, it wasn't to restaurants that depended on the quality of the dining experience. And they were willing to pay for the flavor advantage in the heritage breed. The rush to high-production farming left a small but very lucrative niche for that stubborn old farmer—if only he had lived to see it—and a very pointed rebuttal to Earl Butz.

I put down the paper and picked up the calculator. Six months to maturation, eighteen pounds average weight, three dollars per pound—yes, that was a better profit margin than sheep. The farmer's son was selling the flock. I bought a tom and two hens. Not exactly brave, but I was learning that farming involves managing risk. A tom and two hens could produce sixty turkeys in six months. I didn't even know yet where I would put them all.

Of course, hens won't lay until spring, which was three months away. Okay, so sixty turkeys in nine months. I'm still in.

A week after I bought the turkeys, one of the hens fell over dead. Okay, so thirty turkeys in nine months. Many great enterprises started from humble beginnings. Apple started in a garage.

More importantly, at that moment, these two turkeys offered an option, a scintilla of hope. I had found a niche that could enable a small farmer not just to survive but even thrive. I could offer a choice to the American public that the industrial farms, with their emphasis on production, couldn't. I hadn't exactly dreamed this answer, but I had stayed open to its possibility. Like the blind woman catching a butterfly, you have to wait for it to land and then gently, with both hands, encircle what you can't see.

FINDING LAURA

My father's words were echoing in my ears when I woke that morning. Educate yourself. I wasn't yet a sheep farmer and already I was starting a new venture. After feeding chores, I took myself to the library. I was going to be an educated turkey farmer.

I was elbow deep in poultry periodicals when a thin woman with straight hair and large glasses intercepted me. Sarah, the town librarian, had correctly surmised I was the new farmer in town. She had a problem and was seeking assistance. It seems my neighbor, Laura, had a dental appointment in two days. Sarah needed to get word to Laura that the appointment was confirmed. Laura had been waiting weeks for this appointment and needed to know that it was on—she was expected in two days.

I wasn't sure why this was the librarian's affair or why she couldn't just call this Laura, but I was sure it didn't involve me since I didn't have any neighbors, unless you count timber conglomerates and the US Forest Service in that category.

Sarah paused, giving me a look, and then in a hushed, confidential tone, noted, "Well, yes, but there's also Laura and Leonard."

I relaxed my grip on reality, sensing I was in for another rural life lesson. Seems in 1969, Laura and Leonard's van had its final spasm of death on the road bordering my farm, at the exact moment that Canned Heat's "Going Up the Country" boogie was playing on the radio. This portentous event meant they had been directed to that very spot by cosmic forces. So, they had been living in the woods behind my farm ever since. That's over forty years of subsistence hermitage, earning them a mythological persona of mixed valence in the local community. They had a small but committed cadre of supporters, Sarah being one of them. No one knew where Laura and Leonard lived, exactly, because they moved camps frequently and were, well, socially retiring. Their forest craft was so skillful that they could not be found unless they wanted to be found.

And that was Sarah's dilemma. Laura had a severe tooth ache and had asked to be seen at the local community clinic. A dentist willing to do pro-bono work had been located and would be there in two days. Usually Laura visited the library a few times per week to read, get warm, and stay connected to the millennium. However, she hadn't been seen recently, and time was becoming urgent.

I was sympathetic to Sarah's plight but wasn't sure how I would get a message to these ghost creatures. Sarah paused. "How about I come by early tomorrow and we search together? Just a short hike in the woods." Sure, I thought, I could use some social contact myself.

The morning was cold and damp, with curtains of mist obscuring the contours of the forest. At those moments, it can feel in our dense, coastal woods as though the veil between the physical and spiritual world is lofting open, allowing all manner of creatures to pass through. Sarah selected a promising path, but soon it began closing in on us. We stumbled through underbrush, our feet tripped by vining maple and our clothes snagged by blackberry thorns—or were they ghost fingers pulling us down? We didn't stop to inspect, instead calling louder to Laura and to reassure ourselves with the sound of our own voices. At one point, in a hollow of a tree, we found a cache of poles and black plastic tarp secured with baling twine resourced from the farm. We were initially encouraged that we might be close to their campsite, but it was just a cache. They probably had dozens hidden throughout the forest.

The trail opened into a small clearing. Probably an old campsite, but no sign of residence now. There were paths leading toward the creek, but made by who or what? Going where? When? Impossible to say. The undergrowth gave way to moss and rock and the sound of rushing waters.

Creekside, we came across Laura's handiwork. A pole bridge lashed together with bailing twine, knotted with an economical precision honed from a life in the woods. It was both durable and ephemeral. Sound enough for us to cross but soon it would be overtaken by seasonal floods. It stood as an expression of a life attuned to the transience of nature.

The hour was growing late. Sarah was due back at the library, and my sheep needed feeding. We surrendered to the enormity of the forest and instead wrote Laura a note, wrapped it in a sandwich baggie, and tied it to the railing of the bridge. Then we left.

We never located Laura, but we did find a new friendship. Sarah and I realized the hike had been an adventure—an adventure worth repeating. To this day, I start my mornings with a cup of coffee, a brisk hike, and the companionship of a woman who will trudge the coastal forests to deliver a message of assistance to the less fortunate. That was worth finding.

And after a fashion, we did find Laura. We found her trails and her woods. We smelled the moist earthiness of moss and ferns. We saw mist swirl and envelope her trees, only to be broken by shafts of sunlight striking through to forest floor. We heard the thousand-voiced calls of rocks splashed by water as her creek rushed to the ocean. We found her bridge and her expressions of life in these woods. In the end, was she ever really lost except to us?

And did Laura ever find our note in time to keep the appointment? Sarah didn't know. A week later I went to the barn to toss the morning hay to the sheep. Suspended from a rafter was an origami lamb, made of tufts of wool and strips of woven cedar bark. Attached to the lamb was the note we left at the bridge with "thank you" scratched on the back. Seems I made two friends that day. One from the sunlight and one from the mist.

THE SHEAR OF SHEEP

February is sheep shearing time in our little valley. I didn't know this until Brick announced it one morning over coffee. You want to shear before the ewes are too pregnant to handle but close enough to delivery to count. Our sheep are sheared to make the delivery more hygienic and reduce risks of fly strike, rather than for the wool. We shear in late February for those reasons . . . and because it's when Jesse is available. Professional shearing is not a job that many entertain as a career path. When Jesse is available, we shear.

The first task in shearing is getting the sheep in the barn. They have to be put up overnight so the wool is dry. That involves moving the sheep. Being wiser now, we signed a prenuptial before attempting to move them. Moving sheep is one thing. Getting them in a barn, which requires prodding them through a little door into a dark room, is another. They would stack up at the entrance, refusing to budge. Then one would leap over the entire wool jam to freedom. Then they would all break and run in every direction. Round up and repeat with added emphasis. Around the fourth revolution it just becomes déjà-vu.

This time, with the sheep in stack formation, we decided not to push but to consider our options and out-think them. We conferred. Neither of us had an alternative option. The sheep looked at us. I read willful resistance in their eyes. Greg felt sure it was a surly, mocking reproach. Then one just walked through the door. The others watched for a moment, then stampeded in. We raced to latch the door shut. Yeah, that's how we do it. Out-think them.

Next morning a beat-up farm truck deposited a middle-aged man with stiff joints and a calloused face in front of our barn. He went in, surveyed the barn, came out, and told us to get more lights or forego the shearing. We wasted no time in scrounging enough lights to do surgery in the stall he would use as his work station. He secured long, spring-loaded arms for his electric shearers to the stall wall, then oiled his shearers and checked his supply of replacement blades. This was followed by back stretches to limber up and substituting his shearing moccasins for his boots. By the end of the day, the floor would become so slick with lanolin that only the moccasins would hold a grip. He was ready.

Greg waded into the flock, grabbed a victim by one leg and walked it backwards to Jesse, who flopped it onto its back. Beginning with the belly, Jesse cut the entire fleece off in one piece, flipping the ewe between his legs as he went. When it was done, the ewe was released while I opened the gate. With its modesty assaulted, the naked ewe jumped to its feet and, after collecting what dignity it could, trotted out of the barn. I gathered the fleece into a twelve-foot burlap fleece bag while Jesse was already flopping over his next customer. It went like that all day until the sheep were shorn and the shearer's back was shredded.

It was late afternoon, but the dark barn held the chill and Jesse gladly accepted a hot cup of coffee. While he packed, he reviewed the age and conditions of our ewes, making suggestions for culling to improve the herd. We talked weight and price and which brokers could be trusted and what other farmers were doing to respond to the ever-fickle lamb market. Then he arched an eyebrow and noted that our dark barn made it one of the least desirable stops during his season. Greg took note, knowing that if Jesse dropped us, he would be the likely replacement. A barn rehab just moved to the top of Greg's priority list.

Not all the sheep were sent to the pasture. Romeo went sullenly back to his bachelor's quarters until next breeding season.

The following morning, just looking at all those naked sheep during feeding made me feel cold and irresponsible. I had spent the better part of the year learning to recognize the uniqueness of each of my ewes and now I couldn't tell them apart. Naked is just naked. I consoled myself that the grass was already greening and spring was just around the corner. Of course, then it snowed. School was canceled. We made the news in Portland. The sheep glared at me. I wondered at our luck with farming—shearing sheep and then it snows.

TURKEY EGGS

In March, our turkey hen began laying eggs. When she achieved a sizable clutch, she set on them. A week into it, she fell victim to a coon attack. To our surprise, the tom took over. For the next three weeks he did not budge from his task. Then one morning I heard the persistent chirping of hatchlings. I tried to get a peek, but father tom was having none of it. Each morning brought a larger chorus of peeps. Initially I decided to leave well enough alone, until I found a dead poult pushed out of the nest. After this I decided to be more active in my monitoring.

Over the next few days the mortality rate rose alarmingly. Two chicks disappeared—rats, I assumed. One chick drowned in the watering tank; two others suffocated under the weight of the tom. Time for an intervention. Driving the tom off with a stick, I scooped the remaining poults into a cardboard box. Off to the kitchen infirmary where I could keep a close eye on them while they basked under a heat lamp. A dozen remained. I drew my line in the sand, death would take no more. These chicks were going to be birds.

What I had yet to understand is that turkeys have a will to die. In short order, two went limp and fell under their sibling horde's trampling feet. I scooped them up and stuck them into my bra. After all, a line in the sand is . . . well, kind of scratchy actually. Nestled against my breast, they both revived.

It was at that moment that Greg came in for a water break. From the cubby next to the stove, he noted the box of fuzzy chirps.

He took his time. "So, looks like we're raising turkeys in the kitchen now. I have just one question. Are the birds coming to us or are we going to the birds?"

"Don't worry, it's temporary." I launched into the whole story about the tom with infanticidal tendencies. I almost had Greg convinced that this was rational behavior until a chick popped its head out of my bra. There was a pause. Greg was looking at me with a turkey chick sprouting from my breast. I decided to act busy and started washing dishes in the sink. The chick launched a furious storm of "feed me" peeps. I tried humming but I'm afraid a chirping bra was a touch over the plausibility threshold.

He put his glass in the sink. "Please tell me this is not a new food fad, like grass-fed beef or free-range chicken, because breast-fed turkey breast just doesn't have the same panache."

I hit him. And he's lucky that's all I did.

Still, he was right. We had crossed into some new distortion of reality. The farmer feeding chickens in a cocktail dress, with turkeys in her bra. Well, so be it.

I wasn't there in time for every poult that went down. When the brood dropped to eight, I was in trouble. Without inventory, you don't have a business. That's when I discovered the Neiman-Marcus of poultry catalogs—Murray McMurray. A week later I had a dozen new poults to supplement my declining stock.

The kitchen was warm and bright and convenient for monitoring, but as with all silver clouds there was a dark lining. Mine was a black cat. Bezel repositioned his napping spot atop the turkey cage. Every so often he would rise from feigned slumber and stretch the length of his arms through the bars and wiggle his little paw fingers at the poults. They responded with a chorus of alarmed chirps and sidled, en masse, to the other corner of the cage. Bezel would then hunch up into a pouty scowl waiting for one bird to not notice his intense predatory gaze. Meanwhile, the mice ran free.

It's a funny thing about raising birds in a kitchen. One day everything is fresh and smelling of home-made cookies and the next day it's rancid with bird drop and molting feathers. No longer small, downy balls of wonder, our chicks were now gangly teenagers. And like teens everywhere, it was time to push them out of the kitchen, but not fully into the world. We introduced them into the poultry yard in stages, allowing them a separate space before uniting with the adult birds. Turkeys are very territorial and will attack strangers.

Greg was happy to get his kitchen back. Bezel was initially despondent at the loss of bird hunting options in our house but recovered sufficiently to place the mice on notice. The turkeys grew into silly, beautiful birds—but that's a story for another time. For now, I was still in the turkey business. Humbled in my ambition, modest in my launch, but with sufficient inventory to be commercial. I was farming.

BARN BUILDING

"Morning." Brick was coming through the front door with our newspaper in hand as Greg was rushing off to work. I was pouring coffee. I passed a cup to Brick as he passed it on to Greg.

"Go make us some money." Brick hollered

"Your tax dollars at work." Greg saluted Brick with the cup and was gone.

Brick settled into his mayoral duties, reading the paper while taking account of the goings on of Hopping Frog Road. I told him of Jesse's critique of our dark barn and the fear we could slip from his schedule. This drew Brick's typical response: a toothy grin of anticipation. "Let's go take us a look."

Inside, our eyes slowly adjusted to the darkness of the musty old barn. Built in the 1930s, it stood three stories high with the top two stories designed to warehouse loose hay. The bottom story was divided between milking stations for dairy cattle and lambing stalls. The manure trough was still caked in poop from cattle that had been sold off in the 1970s. The floor sloped to the far right corner. From the outside, it was obvious where gutters had given way, pouring water under the foundation and washing out the rock pillars. So the job of rewiring the barn had just expanded into a new foundation as well. I began to feel the sensation of sinking again and the thought occurred that, maybe, there was merit to leaving the barn dark and me deluded.

And just to put the torpedo to that blissful pursuit of ignorance, Brick added, "With that manure sitting on those timbers for forty years, what you got is a rotting wood floor supported by a foundation of air." I had a three-story building about to fall in on itself and all I could think of was the army of engineers, carpenters, electricians, plumbers, roofers, crane operators, and a catering company that would be necessary to repair it.

Brick smiled again, "Good thing you got me to fix it."

"You? I thought you were a logger, not a carpenter."

"It's all wood. Same principle. Measure once, cut twice, and glue the mistakes. Twelve bucks an hour. Should take two weeks, maybe longer depending on what I find when I open it up."

I still looked skeptical. So he attempted some genuine reassurance. "It was built by loggers, it can be repaired by loggers."

"Okay, let me check with Greg, when can you start?"

"Soon as I get my dog."

I was skeptical again.

"There's bound to be rats under those floor boards. The dog will take care of that."

Well, of course. Or maybe we could go back to leaving it dark and let sleeping rats lie.

Three days later, Brick and his dog had the stalls torn out, the floor boards up, and the barn resting precariously on a network of bottle jacks. This was not pleasant work. When removed, each board proved an archeological treasure trove of calcified cow poop and fossilized varmint skeletons in perpetual repose. The barn itself was laid bare to its bones. The timbers were giant single trees stretching the length of the barn and set on pillars of rock (piers). The pier-and-post foundation made the barn portable. Just hitch a large team of horses and move it to a new location should the need arise, as in the creek flooding. Except, our barn had a sagging timber in the corner. It was punked to the point of cardboard. Brick sank his pocket knife through eighteen inches of wood as easily as opening a box at Christmas.

I looked in disbelief at the dimensions of this log and began to calculate the cost of replacing it with a commercial-grade beam. Of course I had momentarily forgotten our earlier conversation.

From the hay loft, Brick surveyed the forest until he saw the spire of a right-sized cedar. Out of the back of his truck came a chain saw. Thirty minutes later the cedar lay on the forest floor, neatly trimmed into a single log. But a log on the forest floor is a long way from being a timber in a barn foundation. The difference lies in the skill of logging.

As it so happens, Greg had just purchased a new tractor. To be clear, farmers use tractors, loggers use crawlers. Crawlers are short, powerful tractors with low centers of gravity and continuous treads used to skid logs to their intended destination. Tractors sit high off the ground to avoid mud and ground hazards, making them very tippy and not suited for logging. And did I mention, this was a new tractor? More specifically, this was Greg's new tractor.

If you let little things, like reality, intrude on your plans, you're probably not cut out to be a logger. And Brick is a logger's logger. Up into the woods Brick went on Greg's shiny new tractor. He chained the log to the front end loader, estimating just the right fulcrum to tip the log completely off the ground and, carrying his trophy like a leaf-cutter ant, chugged off the mountain. A slight shift or misguided bump could have toppled the tractor into a fatal roll down the mountain. At the bottom, he glided the tractor around in a smooth arc and lowered the log into place like a banker counting money. Across his face was a large, self-pleased grin. Brick was logging and loving it.

I stared in amazement. An hour ago that log had been a tree in the forest. With a little luck, it will still be supporting my barn a hundred years from now. Brick hopped down and detached the chains, then used the loader to fine-tune the alignment of the log. Then he used spikes to nail it in place. And that was it. He sat on the beam, patting his dog stretched across his lap.

"Tomorrow I'm going to fix the piers. After that we're going to need lumber to replace the rotted floor boards. Should take a unit and a half of rough cut. They don't sell that at Home Depot. I'll call over at the mill, but you're going to need someone to pick it up. Maybe Will Ridout's hay truck could haul it."

The tractor was parked back in the shed without a scratch. Until writing this now, I don't think Greg knew how close his tractor came to being scrap. Probably it never came close at all, given that Brick was a gifted logger. What he wasn't, though, was a carpenter. He could tear floor boards out, but putting them back would require Luther and about four more weeks. And what I didn't know was that the steelhead run started in a week, so I was about to lose one rambunctious logger.

Like everything else in the country, schedules are organic. Commitments do not adhere to the tight structure of the factory, as they do in the city, but to the ebb and flow of nature. And, in Oregon, commitments are especially malleable when the fickle inclinations of those silver fish drive them upstream.

LUTHER AND OTHERS

Brick prepared his exit by referring us to Luther, who he promised was a skilled carpenter capable of speeding things up. I had heard stories of Luther. A lifelong bachelor, he lived with his dog in a bunkhouse behind his mother's place just off the main road. Luther tended a small farm, supplying most of the fresh vegetables at the Merc. He did a little farming and the occasional odd job, but mostly Luther kept to himself. It was rumored that he ran a still up in the woods and that he consumed more than he sold. Some thought him a ne'er-do-well, while others thought he just about had it all figured out. One thing was certain: he lived life at his own speed, without the slightest concern for what others thought. The only schedule he recognized was the cycle of the moon that guided his planting.

The stories did not match the breezy attitude of the man ambling up to my door. Blue eyes, straw hair, sun-burnished face, he sported huarache sandals and faded blue jeans held up with a multi-colored Mexican peasant sash. The look was more gypsy surfer than hillbilly hermit. And that smile, like a neon sign flashing "wanna ditch school?" I wondered how many women had reclaimed their inner schoolgirl while cruising on the bench seat of his beater pick-up truck.

I opened the door and he held out a widget—a trellis made of willow and split cedar. It was part garden furniture and part folk art.

"It's my calling card. Keep it and think of me, if you need any handy work."

Greg stepped out from behind me and examined it. "I'm impressed. What kind of work?"

"Two hands, two feet kind of work. Whatever you got. I can usually figure something out, but I like working with wood."

Now as a point of fact, I had neglected to tell Greg that Brick had referred Luther. Just as I was about to correct that omission, Greg veered the conversation into the fast lane.

"I need a bridge over the creek. We got a footbridge, but it washes out."

"I can do that."

"But I don't have any money."

"I can't do that."

Before we went too far down the wrong path, I intervened. "Umm, I think he's here about the barn . . . And what bridge!? We don't need a bridge."

"'We don't need a bridge?' How do you plan to graze the far field when the floods hit? And what barn?! You mean Brick's not finished stringing lights?"

"Well, I may have forgotten to mention, it turned into a little more than stringing lights."

"What?! How much more?"

"Well . . ." I said with exaggerated impatience. "That's why Brick's friend is here."

The man with straw hair and huarache sandals looked at us both and offered, "I got some pot in the truck if that will help?" Then he smiled and added, "'cause I'm starting to feel a little stoned just listening to the two of you."

Well, that fit. A ditch-school smile and pot in the truck. I knew Greg was now thinking he wanted this guy off his property. Before he could serve his eviction, the gypsy surfer continued, "I like to think I'm friends with most people up here, but which one sent me?"

They both looked at me, expecting illumination.

"Aren't you Luther?"

"Nope, and I checked before I left home. I'm Jack. I bet you mean Luther Vinson. Lives off the state route. You got him signed up to do your barn? Well, he does good work, least he did the last I knew.

"I've been gone ten years. Just moved back. I was up on Whigby Island working as a handy man for rich people. But my daughter called saying she needed help with expenses. We've always been close, so I moved back to help her on her farm. Well, that, and my Catholic girlfriend converted to witch. She said my male energy was creating a ripple in her vortex and

I should move outside. She packed my stuff and put it in the truck. When I released the parking brake, it just rolled right on down to my daughter's place. So I'm back and looking for work but looks like I'm too late for this job. So, Brick who?"

"Brick Reahl."

"You mean Rick. You got Ricky working on your barn? I was his shop teacher in high school. Ricky is a prankster. Locked me out of the school bus on a field trip. Course he forgot I had the keys and enjoy a good prank myself. He missed supper that night. Ricky is a good man, but Luther will get the job done.

"Well, you got my card. Let me know if you want to do the bridge. I got thirty dollars an hour on Whigby but here I'll do it for ten."

Apparently, Jack knew how to get to "yes" with my husband.

"Ten dollars an hour?"

"Yup."

"It's about a forty-foot span and needs to support the weight of horses."

"That's a significant bridge. Tell you what, we can harvest the cross beams on your property. That will save you enough to pay my labor. But I'll need your help setting the beams. What do you say? Wanna build a bridge?"

"Just one thing. I can't have drugs on my farm. Can't afford the legal entanglements."

Jack looked confused and then remembered, "Oh you mean the pot. That's barter—better than money. In these woods, you'll find it opens more doors and closes more deals than greenbacks. But your farm, your rules. No drugs."

Greg nodded and they shook on it. It was going so well, I thought maybe I could add on, "And maybe when the bridge is finished, Jack could help with the barn?"

Jack drew back. "That's a little tricky. A fella piggybacks on another fella's job and that fella might start to feel a little stepped on when he sees his paycheck nibbled down. Probably best if we leave all the stallions in their stalls."

Okay, then. No piggybacks. Only stallions in stalls. I couldn't help but wonder if it were women building a bridge, wouldn't they welcome the help? Maybe not. Mares need stalls too. Still, I liked thinking about the stallions in their stalls.

Two weeks later, Greg had a substantial bridge over our creek. It was the first "substantial" thing he had ever built, and a great source of pride. More than a carpenter, Jack was also a teacher. And more than a teacher, Jack was also a sculptor of the social vibe. A true artist of folk craft, he used wood and music and story to bring folk together. But we didn't know that yet. Greg just knew he liked hanging out with Jack and they built a helluva bridge together.

THE HERMIT AND THE GYPSY

Luther did eventually arrive, looking like a wooden mannequin occupying the passenger's side of Brick's truck. Dressed in T-shirt, farm overalls, boots, and red bandana enclosing long, graying hair, he offered no invitation for engagement except a full, two-syllable "Hell-lo" that initially rose in cadence, then dropped in the dirt—seemingly exhausted under the weight of so much social obligation.

We walked over to the barn to survey the scope of the project. From the distance came the rhythmic pounding of nails. I noted that construction on the bridge was underway as a simple point of reference. Luther could not have cared less. Brick pointed to the unit of rough cut lumber and asked if we wanted the more expensive galvanized nails used on it. I deferred to Luther's judgment.

Luther responded that it wasn't his barn, and wasn't his concern if we wanted to use cheap-ass nails that would rust out or not.

So galvanized nails it was, then.

As for start date, Brick noted they were here now. Good because we needed the barn finished in time to put up the hay. They both nodded and began to square themselves away on job priorities. I was sent to town with a list of supplies, including galvanized nails.

Before leaving, I noted Greg would be able to assist as soon as the bridge was complete. Brick, looking sheepish, responded, "Good, 'cause I can give you two more days and then I have another commitment." *Commitment* being code for steelhead run on the Siletz River.

I looked at Brick and then at the cavern that formerly was a barn and then at Brick again.

Brick rushed to reassure, "Luther can finish up. Truth be told, Luther's twice the carpenter I am."

Luther looked at me, waiting to see if Brick had just canceled their work order. I asked point blank, "Can you finish without Brick's help?"

"Always have."

Apparently this has happened before. "Well, guess I better get those supplies so I don't hold you up." I shot Brick a wilting look that he pretended not to notice. He was practiced at escaping the judgment of women.

They put in two full days of work. The third day, Luther's mother dropped him off in the morning. The fourth day, she dropped him off at noon. The fifth day she went into town and he missed his ride altogether. A pattern was emerging.

I was inclined to fire Luther and hire Jack, who had been prompt, and true to his word. But this is a small town and firing a member of the community could have unforeseen ramifications. Asking them to work together also was fraught with complications, as Jack had indicated. Greg had a solution. Hire them separately. Give each stallion his own stall, or in this case, own side of the barn to work on. The competition might improve Luther's work ethic. Greg would function in a swing position, helping both and making sure the competition remained friendly. We would start fresh on Monday.

Monday morning, Jack showed, eager and willing. Luther was absent. We called and Luther's mom walked the phone to the bunkhouse. Luther was still in his bath. He would be on the job as soon as he finished his soak. When would that be? When he finished.

That tore it for me, but Greg wanted to give Luther one more chance. As Greg explained to Jack, he hoped competition would bring out the best in Luther.

Jack smiled, "Luther's not into team sports." Then he winked and climbed into his truck, "Honey might work with flies but not so much with cats. Leave it to me."

An hour later Luther was on the job, looking freshly washed and as close to motivated as Luther ever got. Jack had access to motivational aides that were not in Greg's toolbox. As Jack had noted, money only goes so far in the country, especially with asocial, self-sustaining misanthropes like Luther.

Jack asked Luther where he wanted him to start. By handing Luther authority, Jack neatly avoided conflict. Luther took a few measurements and then directed Jack to start cutting boards to his specifications. Greg would transport the boards to Luther, who would be nailing in place. The construction was underway. Everyone found their role in service to the overall goal, submerging the struggle for dominance to the quest for accomplishment.

Gradually the tedium of pounding nails was interspersed with stories of small triumphs, painful losses, and acts of defiance against The Machine—meaning, whatever form oppression took in their respective lives. Jack initiated, but each man found in the others' stories a portal into their own. And as they caught up on each other's lives, they found common themes that opened new portals to new stories that connected them in felt experience just as the work was connecting them to the project of building a tangible barn. By day's end, tired bodies acknowledged a day of accomplishment that made the pain worthwhile and sustained the men's spirit. The stories had woven them together in a cloth of companionship, and the weaver's hand was so deft that it was barely seen.

Jack drove Luther home and arranged to pick him up in the morning——right about breakfast time. Seems Luther's mother made the best cinnamon buns in the county. Luther was never late again—or not as late—and Jack was never hungry again.

The men found their woodworking skills to be complementary. Luther's slow, methodical style curtailed Jack's tendency toward impulsiveness and reduced errors, while Jack's enthusiasm kept Luther on task and moving forward.

And that's how a gypsy surfer and a hillbilly hermit became the Wayfaring Brothers Construction Company. No job too big or too small, depending on the whether . . . meaning, whether they're in the mood.

PEEPING TOMS

Chickens can't fly, but they can manage a pretty good airborne hop. A six-foot fence will keep most chickens cooped. Turkeys will fly the coop. Heritage turkeys can catch enough air to sail over any fence a farmer can put up. Fortunately, they sail back at dusk. By mid-summer, our turkeys began free ranging beyond the coop. The dogs weren't having any of it and chased them back. Thus began one of those epic struggles between freedom and limits. When the dogs were away, the turkeys tasted freedom. And for turkeys, freedom tasted a lot like fresh grubs and grapes.

Through the summer the turkeys alternated between harassing the chickens, scaring the cat, running from the dogs, and hanging out with us while we did our chores. We would look up and find them staring at us through fence rails, or dodging sprinklers, or begging for lunch scraps from the farm help. While the hens were timid and uniform in personality, the toms were adventurous and provocative, with displays of tail feathers declaring their uniqueness. What started as strutting often escalated into tussling and squabbling matches with lots of commotion. And when they weren't showing off, they were running in packs patrolling their territory. One poor delivery man was cornered by a circling gang of gobblers in full display. Cut off halfway between his van and our house, he was frozen with package in hand. "Beady-eyed devils" was his description of our heritage bronzes.

They were great at sound mimicking. Any discordant noise would draw a chorus of gobbles. Greg took to baiting them, calling to them in the morning and waiting for a response. He tried a rhythmic, turkey rap repartee with some success until complaints from the neighbors shut it down. Art ahead of its time.

Of all their antics, the toms were at their turkey best in the summer evenings. We have large, glass doors that open to a back patio. Regularly of an evening, three to four toms would be posted at those doors peeping in. Their long necks craning to see what lay beyond those panes of glass. Then they would bob and weave, looking for a better vantage point only to return to their tippy toe, full stretch stare. When their frustration reached a peak, they would start pecking at the door. This brought snarling dogs charging at them, causing them to scatter to the four corners.

I assume they were responding to the turkeys they saw reflected in the glass, but who can say what goes through those pea brains programmed by mother nature? I do know that having large birds doing the bob and weave shuffle on your door step is both a delight and a mystery. And far more entertaining than anything programmed on the TV.

In November, the truck came to collect the turkeys for market. I had mixed emotions. Loading lambs into an unfamiliar van is a nightmare. Their instinctive response is to protest, which seems a natural response to the impending event. In contrast, the turkeys allowed us to scoop them up without a fuss and place them in the van. They stood huddled in a tight group, wary of the new surroundings, but making no attempt to escape. Their vulnerability and their utter trust of me made it feel like an even deeper betrayal than with the lambs. It was just so sad. After that day, I swore I would never raise turkeys again. Maybe I wasn't cut out to be a farmer—at least not a turkey farmer.

That is, until our butcher mentioned how many pre-orders he had for next year. His customers raved about the taste. But then, they were used to commercial turkeys. These turkeys were heritage. A breed that existed only because a small farmer refused to "go big." And they were raised on a small farm where they had a good life. They were beautiful, healthy birds that wandered our farm, digging grubs, chasing cats, and doing their turkey dance to their heart's content. They had provided the local schoolchildren with an up-close turkey experience that produced an abundance of questions (what's the thing on their nose called? a snood) and colorful drawings. They had a happy life and they provided nourishment for us.

Are we justified in taking another life to provide for our own? I believe so, if it's done as ethically and humanely as possible. I am proud that my animals have a good life; I don't think I will ever lose the sadness I feel for their sacrifice, though, and I've come to realize that's a good thing. The sadness comes from caring about the quality of their life.

So, we will be raising turkeys again. But I swear, those chicks need to make it on their own this year. And like most of my oaths, this is iron-clad . . . until it isn't. Until I find a chick, fallen from the nest and struggling to survive. On our farm, the circle of life often involves a layover in the kitchen infirmary.

PART THREE

ANOTHER NEW YEAR OLDER

The snow fell, separating us from all our tribulations. Our girls came home and we celebrated another small Christmas. There was no disappointment this time because there was no expectation for more. Less was not only enough, it was comfortable, it was intimate. We lost power and huddled together under lamplight, warmed by the woodstove, enjoying the respite from our chores. We were family and we were community enough for each other to lift the dark of those December nights.

The New Year brought the annual tally of income and expenses. If it was like last year, it would be followed with grim atonement for the sins of ignorance and naiveté.

The net from the sale of the lambs was up 5 percent. It was the right direction, showing we were improving, but not nearly enough. The net from the turkey sales doubled the net from the lambs. Now that was more like it. And we reduced our feed costs by haying. Unfortunately, all the gains were off-set by the monthly tractor payment. In five years we'd have the tractor paid and then the gains would mean something. It would mean we would be losing less money.

Farming is a high capitalization business and it always has been. Even when the government gave land away for homesteading, many could not "farm up." They lacked the capital to buy the necessary tools, seeds, and livestock to farm. Today, with four-hundred-thousand-dollar combines sitting ten months of the year in tractor sheds, the cost to farm-up has never been higher.

This is the point where rational people look at the numbers and acknowledge it's been a great adventure, but it's time to quit. Of course, rational people don't leave lucrative careers to buy money-sucking farms in the first place. Greg's income was enough to keep us afloat, and he found he actually enjoyed the variety offered by the dual career lifestyle. One occupation was nearly the opposite of the other in skills and activity, which kept him fresh and engaged. As for me, the principal farm operator, I still felt I could make the farm pay for itself. As I noted, rational people don't buy farms.

With small farms, it's about the niche. That's what the turkey sales were telling me. I had to find those areas that big ag couldn't or wouldn't address and farm to that market. So, it's not about one niche, it's about finding multiple niches. In that way, small farming differs dramatically from large scale, monoculture farming with ever greater levels of one-crop efficiency. Small farming is about maximizing diversified crops synchronized to the seasons, enabling the small farmer to be nimble and responsive to the market. How could I apply niche marketing to sheep?

After researching sheep breeds, I hit upon Katahdins, a hair sheep from Maine. Hair sheep shed their coats, saving me the cost of shearing. Remember, sheep are bred to specific purposes, so the wool from meat sheep is unsuitable for textile. Coming from Maine, the breed seemed well adapted to Oregon wet, reducing the risk of hoof rot. Further, they were much leaner than other breeds, imparting a milder, more delicate flavor to the meat. The flavor advantage should command a premium price, but only for a consumer educated enough to ask for Katahdin. So how was I going to find—or develop—that educated consumer? Not likely the big grocery chains would do it. No, I needed a group of consumers I could talk to directly. I needed a community of consumers that supported local agriculture—a CSA (community supported agriculture). So my problem was two-fold: find a breeding herd of Katahdin, and find a paying group of consumers. Easy peasy.

Not every advancement could be counted in dollars. This year we had greatly extended our connections into the community. Greg had joined the Grange, a century-old fraternity for farmers to connect and exchange ideas. I was mining the Ag Extension Service for workshops on farming. The Extension Service is pure Jeffersonian agrarianism, linking farmers to ag scientists at the land grant colleges. It keeps farmers tuned up on the latest developments in science, and the scientists grounded in the practical needs of farmers. And that keeps America at the forefront of farming. Small farmers don't get government subsidies, but we do get the Extension Service, and that's a lifeline in a stormy sea.

My agent at the Extension Service was Melissa. She was part of the Small Farms program, which was responsible for answering questions, registering concerns, networking, and providing educational classes. She connected me to a group of women who farm. Not surprising, since among small farmers, women managers are a growing demographic. And that has had a variety of subtle effects on the cultivation and marketing of food. Our Farmer Jane potlucks are always tasty farm testimonials.

As for our marriage, it was fine. Without fully discussing it, we had divided our labors and were now engaged in their execution. Greg provided the financial support through his job and tended to the garden and pasture cultivation in the summer. I tended the animals and much of the business of running the farm. We had each gained in competency, and that, combined with a better farming strategy (farm the niche), was improving our financial outlook.

Or to frame it another way, we were completely dependent on each other without offering a jigger's worth of support to the other. If either of us had let down for a moment, the whole venture would have collapsed. We knew this abstractly, and we knew that the other person was working hard, but the separateness of our lives prevented us from seeing the exact work and simply acknowledging it or the person producing it. We saw our partner's effort only when they failed, and then we brought that failure to their attention—fortunately, this was not often. In short, we were desperate, dependent, and alone in our struggle to support each other—and doing a good job of it. So, we were fine, thanks for asking.

MORE PEASY THAN EASY

Applying niche marketing to sheep required solving two problems. First, finding a distinctive brand of sheep that would command a premium price and then growing—literally—that brand. Second, finding a dedicated community of consumers who were willing to pay a premium and growing that community of consumers. There are 1,200 breeds of sheep, so selecting a brand is not an easy process.

The story of sheep is the story of us. Sheep are easy pickings for predators and probably would not last long as a species without the protection of their chief predator. It's a symbiotic relationship that began ten thousand years ago in the Levant, making sheep the first livestock. They were probably domesticated for their milk, not meat, suggested by the absence of butchering sites in their place of origin. Sheep's milk, with its high butterfat, is still the best dairy source for quality cheeses in the world. In my mind, I envision some poor man who's just lost his wife and is desperate to feed his newborn, cornering a feral sheep and milking it. Viola, in a few short millennia we're enjoying a piquant Roquefort at a Paris bistro. From this initial contact, it would be three thousand years of selective breeding to produce wool sheep. It would be another three thousand years before Jason would go in search of a Golden Fleece or Abraham would move his flocks across the desert to Canaan. Sheep have shared our journey and earned a place in our stories since antiquity.

My choice of sheep was Katahdin, a hair sheep derived from the St Croix breed. St Croix were brought with slaves to the West Indies, suggesting an African origination of the breed. Like most hair sheep, St Croix are incredibly tough, resistant to parasites and disease, and able to survive on marginal lands. Katahdin have inherited all those traits along with extra resilience for cold, wet weather climes like Maine and Oregon. The one drawback is that they are smaller and leaner by a third than the traditional woolies. When profit is measured in pounds, that's a problem. Of course, my intent was to flip that formula so profit is measured in flavor, where the Katahdin contain that same proportion of advantage compared to the heavy, tallow flavor imparted by the fatty meat of the woolies.

I had found my brand of sheep, now all I had to do was find the actual animals. That required a call to the national Katahdin sheep growers association for a referral to the nearest local growers. Then multiple calls to local growers to find the right quality (birth/weight record), availability, and price for breeding ewes. That could take several months, and several months could cost me a year of product if I missed my breeding window.

Or, instead, a call to Melissa, who knew a woman in a neighboring state selling her entire herd of Katahdin breeding ewes. A week later I had my breeding flock.

First look was not encouraging. The ewes were shedding hair in clumps, giving the distinct impression of mange. They were small and feral looking, almost like miniature, pot-bellied deer. And they came in assorted shades of beige, not the pristine white I associated with sheep. When introduced into the larger flock, they immediately separated into their own clan. The overall impression—small, feral, clannish—was vaguely reminiscent of my Mainer cousins. Fortunately, they turned out to be just as endearing as my Mainer cousins.

With the breeding herd secured, all that remained was culling the old herd and finding a ram. Culling was necessary to prevent over-grazing. Grass does not grow evenly, but rather follows an S curve rate of growth, meaning it grows slowly at the bottom and top of the curve. Efficient grass production requires four inches of grass to maintain maximal growth. Grass cut below that baseline will be stunted. The herd must be culled before the pasture is over-grazed.

Culling involves reviewing lambing records. It's a matter of eliminating the oldest and least productive. Once the cull is made, those ewes are taken to auction. At the auction are a group of professional buyers who will assess the animals relative to current market conditions and offer competitive bids. Some animals will be sold directly to slaughter while others will be taken to feeder operations. For the buyer, it means bidding, loading, transporting, and selling the animals in the same day, hoping you've bid shrewdly enough to make a profit. For the farmer, it is always the least money you can get for your animal. For a conscientious farmer, it is also the least caring manner for disposing of your animal. Though necessary, it is never a pleasant event.

On the other side of the livestock fence was my purchase of a purebred Katahdin ram. The quality of your herd is determined by the quality of your ram. For this reason, purebred rams command a premium. "Red" was purchased from a biohazard-free farm. It was a closed farm, so outside vehicles were not permitted beyond the designated parking lot. Once out of my truck, I was instructed to wash my hands with an industrial sanitizer and place paper booties over my shoes. Red weighed two hundred pounds, which is big for a Katahdin, and sported a massive mane of red tinted hair around his head and neck, giving him the appearance of a lion. His super-sized testicles added to his lionized stature. Fortunately, his temperament tended more to the lamb. I slipped a halter over him, and loaded him into the back of the truck.

So, first problem solved. I now had a breeding herd of Katahdin sheep. And a few extra, since I couldn't part with several of my favorite woolie ewes. The woolies would go on lambing for another six years and their offspring would keep woolie genes circulating in the herd long after that. By happy circumstance, the cross created a weight advantage without noticeably affecting the meat quality. Unfortunately the cross also created a pitiful looking, half-wool, half-hair creature with a perpetually crumpled look. But that look fit the farm. In fact, both the farmers had that look most of the time. Another example of perfect imperfection.

At this point, all I had done is to transition my breeding herd from large sheep to small sheep. This is not a path to prosperity unless I also convert my consumers to the flavor advantage club. Before that, I needed to find my consumer.

I advertised my Katahdins on craigslist, but since the public doesn't recognize the Katahdin brand, I was forced to sell at market price. I was selling at a loss to find my customer with the hope of building a base of customers that would generate a buzz. I launched a website, describing Katahdins, telling my story, and offering recipes to help educate and network my customer base. Once there was sufficient interest about the Katahdin brand, I could start raising prices to a product-sustaining level. It was a process. And since the kill-cut-wrap price (yes, it's one word in the industry) is the same, regardless of weight, my smaller lambs would always be at a price disadvantage. Meaning, it's a slow process.

So, the Katahdin niche market would improve over time as my consumers become educated. Now I just needed my loan agent at the bank to agree with the slow improvement and notch his interest rate to correspond to my slow improvement rate—which would never happen. And that's farming. That's how difficult, risky, and expensive a small change in farming practice is to initiate and why the deck is stacked against the small farmer. Yet, small farmers are pulling off these odds-breaking miracles every day. Ever enjoy a glass of world-class Oregon pinot noir? The same story of slow education, stubborn persistence, and bone-crushing finance brought that bottle to your table.

AN INCONVENIENT TRAP

Sarah and I walk every morning, whether rain, fog, cold, or clear. We walk to loosen rusted joints and clear our brains. And we walk to stay connected, to sympathize, and synchronize with our families and our community. And I walk for the dogs. Patches and Cisco live for the morning walk. My hand on the front door is the signal to let the party begin. Each dog begins barking and turning circles in their unique version of the happy dance.

I have never understood the dogs' level of excitement. Perhaps I would if they were kept inside all day, but these dogs have their own door. They have unfettered access to chase chipmunks, flush birds, bark at passing cars, and surreptitiously herd sheep. And what do they do with this freedom? Mostly lie around the house and sleep. But look like you're going for a simple walk and these dogs become apoplectic with pleasure. Must be archetypal memories of the wolf pack at hunt.

This day we chose an old logging road with soft grass, fallen pine needles, and no trucks. We call the road the "boneyard" because locals favor it for dumping the offal from their hunts. Either that or a satanic cult is operating in the area with a neurotic need for overachievement. We also call it the boneyard because it is eerily quiet. Something about the presence of death makes even the birds avoid it.

I think it was the quiet that made the shriek coming from the woods so unsettling. A dog was crying out in agony. And the shriek didn't stop. I whistled. Patches emerged from a thicket but no Cisco. I ran into the dark woods toward the terrible noise. If it was a porcupine encounter, the wail should have stopped by now. Knowing deer and elk carcasses draw a variety of predators to scavenge, my mind was filled with dire images of Cisco being attacked by a cougar

Through the gloom of the dense forest, I spotted him at the base of a large snag. I called to Cisco but he didn't run toward me. Something had hold of him. I called again and this time his head twisted in my direction. Concealed behind a snag, much of his body appeared immobilized. As I got closer I could tell his paw was caught. I crouched beside him, throwing off my gloves and reached for his paw. Cisco was in a panic, alternating growling, snapping, and crying. I realized he was caught in a leg-hold trap. The harder he pulled, the worse it got.

I threw my arms around the dog to stop him from pulling while I tried to press down on the trap with my boot. I could barely make it move, and by pressing on just one side, it pinched Cisco's paw harder. He yelped and snapped at me. I was not strong enough to hold the dog and open the trap at the same time.

I yelled for Sarah. When she arrived, I gripped Cisco's muzzle as she threw her coat over his head to protect us both from his snapping teeth. I held him in place while she, with all of her strength, tried unsuccessfully to pry the jaws of the trap apart.

Neither of us had experience with traps. Then, I noticed the trap had spring releases on either side of the jaws. Sarah used her weight and stood on the releases. The jaws opened. Cisco yanked his paw free just as the trap snapped shut again.

I gratefully released Cisco and stood up on shaking legs. Cisco sat down to lick his paw and inspect it. After a cursory washing, he rose and ran down the path. There was no limping or three-legged hopping. He was fine and he wanted everyone to know it. No need for a vet or any further inspection of his foot. Everything was alright. Nothing to see here. He went straight home without a look back.

I tried to thank Sarah but soon realized I was babbling. Like Cisco, I needed to go home, to sit down, and to calm down. Sarah hugged me and told me she was available if I needed her. I don't remember saying good-bye, although I'm sure I did.

As I walked back, I was surprised that my hands had become suddenly cold. I had put my gloves on but my right hand was starting to tingle, so I took them off. My hand was caked in dirt mixed with blood. Curious, since the trap had not broken Cisco's skin. I hadn't noticed any blood on Cisco. That was because I was the one who was bleeding. Cisco had bitten me.

Two weeks later, my hand was still bruised with little teeth marks. The antibiotic worked wonders on the infection. As for Cisco, the vet was amazed there were no broken bones. However, getting to look at his paw required a muzzle, a blanket, and three vet assistants. As for the trap, Sarah and I went back the next day to retrieve it. It took both of us to pull the three-foot stake holding it out of the ground.

I called our local Fish and Game guy to report the heinous crime and found the trap was legally set and registered. Anyone can trap on public lands as long as they register with the authorities. He did inform me that taking someone's legally registered trap was a crime. I asked if he had spelled my name correctly and proceeded to give him the name of my high school algebra teacher.

The trapper came by a few days later to reclaim his trap. He told me he had pulled all the traps from our road. He hadn't realized people were walking with dogs in the area, and he noted, it wasn't his intention to catch family pets. He was a genial man, nothing like the backwoods schizoid my mind had conjured. He was a second generation trapper paid by timber companies to thin the beaver population and curtail the loss of valuable trees. We can debate the merits of a national policy that allows the trapping of beaver in the twenty-first century, but it's not fair to hold an individual trapper responsible for doing what he is legally entitled to do. He was just a guy making a living in the woods.

So who has the right here? The lady walking her dogs, the guy trapping for a living, or the beaver just doing what beaver do? Each in our own way are looking for that balance between the ways of nature and the ways of mankind.

LAMBING

I was expecting lambs any day. The first hints of spring had warmed the earth, turning the grass a luxuriant shade of green and beckoning the lambs to come forth. As for my part, I had read everything I could on lambing. Again, I turned to Melissa, my Ag Extension agent, pestering her with lambing minutiae. I even graduated from the Extension's lambing class with honors. I was prepared. So where were the lambs? I checked and rechecked the gestation charts. The weather was perfect. I was prepared. The time was right. It made no sense, unless those ewes were purposefully holding back out of pure sheep malevolence.

Then it rained. A cold, drowning rain that went on for days, sliding the farm back into winter. Mud replaced grass. The casual walk to the barn became a boot-sucking slog. The sheep scattered to the far corners of the farm. Now the lambs came.

The first ewe gave triplets. When she didn't come in for the evening feeding, I went in search. There she was in the farthest field with the sun setting and the rain pouring. Less than an hour old, cold, and wet, the lambs were at risk for hypothermia. I slid one into my coat and zipped it up to its head. The others I carried wriggling in each arm while mom zig-zagged between my feet with incessant calls to her babes. Several times I nearly slipped in the mud and the dark and the wriggling confusion. Back in the barn, I dropped the lambs into a stall with fresh hay and a heat lamp. Mom rushed through and in no time the lambs were taking their turns at the teat. A good sign.

Lambing is a little like a high school dance. Once that first awkward moment is broken, everyone joins in. I didn't need any more night plods through the mud, so I kept the herd close to the barn. Several more ewes popped . . . and then the complications arrived in waves.

Looking back, you don't remember the easy births, only that there were some. I exhale with relief whenever I count strong babies holding close to their mother's side. Strong enough to stand within minutes. Nearly impossible to catch after forty-eight hours. Which means, better to catch them right away so they can be weighed, tagged, iodined on their umbilical cord, given a shot of BoSe, tails docked, and castrated when gender necessitates.

It's the complications you remember. I lost my first ewe to a hernia. The large fetuses pressed against her internal organs, rupturing the hernia and killing both the ewe and her lambs. Then there were multiple births—triplets or quads so small and frail. Often they would fail within the first few hours. It was harder when they lasted several days. I interceded with tube feedings—milking their mothers for the all-important colostrum that would build their immunities. Placing them on beds of straw, under heat lambs, caressing them to encourage their will to live. Sometimes they did.

There was Poodle, named for the tight, white curls of her coat. She was the runt, pushed from the spigot by her three bigger sibs. I tethered the ewe and gave the lamb unfettered access to the teat but she could barely reach and her sucking response was weak. So I milked the ewe and bottle fed Poodle, leaving her for the night. The next morning she was too weak to stand and her mouth was cold, signs of impending death. I brought her into the warm kitchen, placing her in a box with blankets and heat lamp. I bottle fed throughout the day and into the night. She began to perk up. The next morning, when I snuck down early to check on her, she was dead.

Then there was the lamb I tubed and bottle fed and worried over until at last he was strong enough to hold fast to his mother teat, even while competing with his twin. Still looking healthy after a few days, I released the three to the pasture only to find the lamb weak and wobbly again one morning. I brought the ewe and her lambs back in the barn, supplementing the little guy with bottle feedings. His mouth seemed wired shut and he was uncomfortable when I placed a nipple dripping warm milk against his lips. Clear signs of tetanus, a disease promising an agonizing death, with no hope of survival. I laid him across my lap, rubbing his head and speaking softly, "I've got this." Then I euthanized him. A sad end to a short life.

And the others that failed. Some never took a breath, despite brisk rubbing or swinging their limp bodies in a circle to force air into nonfunctioning lungs. Some were breach (backward) or had become tangled with a sibling in the birth canal, necessitating an intervention on my part, with an arm up the uterus, to shift the bodies until they could be delivered. There were times I was too late, causing them to suffocate in the womb. And, some were just abandoned by their mothers, rejected for a myriad of reasons that only the ewe knew. Often these abandoned lambs, called bummers, died shortly after being abandoned, even though they received all necessary supplements. Somehow the ewe knew the bummer wasn't going to make it and conserve her precious resources for the lambs that would live.

But some bummers are abandoned because the ewe is new to mothering and lacks good maternal instincts. One particularly wild new mother lambed in a secreted corner of the paddock in the dark of night. By the time I discovered her, both mom and babes were bronco beyond catching. Then I found an additional lamb prostrate, with head drooped over his back. His mouth was cold, so I hurriedly carried him back to my kitchen. I readied a bottle of powdered milk replacement. Like most lambs, he had no desire to suck on a synthetic rubber nipple until he tasted the warm, sweet milk trickle down his throat. Then he inhaled four ounces of the stuff, swelling his little belly. Next, he was ready for a nap in the straw. I had fashioned bedding in a cardboard box placed next to the stove and strung a heat lamp above it.

He was an ugly little bummer, and I wondered if that had anything to do with his mother's rejection. His skin shone pink through a light dusting of wool. His legs were too long and his head too small for that round belly. He resembled a freshly plucked turkey leg. I took to calling him Turkey. Greg called him the hairless wonder. None of these names stuck, so in the end, he was just Bummer, the bald lamb.

His health improved and within a day he was hopping out of his box. There was no returning him to his mother; a ewe's rejection is absolute. He was ours now. Like most newborns, he required feeding every four hours. I wasn't keen on traipsing to the barn in the middle of the night for his feeding, so I fashioned a diaper out of a plastic grocery bag and set him back in his box in the kitchen. That didn't last long. Soon as the lights went out we heard the tapping of tiny hooves across wood floors and then the final leap into our bed. He had already introduced himself to the in-house livestock when he nestled between them around the wood stove. Now, with muffled groans, they just shifted on the bed to make room for him.

As I drifted off to sleep, I wondered what those tough, Basque shepherds would think if they saw me now. On second thought, I doubt they got out of their wagons in the middle of the night and walked barefoot through the rocks to feed their bummers. Nope. I bet shepherd, dogs, and lambs all curled up together just like we were doing.

The bummer took to Greg and would lie at his feet in our parlor. One night, Greg announced his name to be "Snickers." When asked where it came from, Greg swears the lamb told him. And it did fit. Everyone nodded approval. There was something special about this little orphan, and Snickers captured it.

This episode lasted a month. Gradually lambs wean themselves from milk to grass. For bummers, they no longer need humans and develop a preference for the company of their own kind. Still it's good to have a few bummer ewes in a flock because they retain a people-friendly attitude that helps tame the herd.

But Snickers, being a wether (castrated male), presented a problem. There's not much use for wethers except as meat. And no one was eating Snickers, according to my husband.

"So we're keeping him as a pet?" I asked a little pointedly.

"No, we're sheep farmers, so no sheep as pets."

My husband has principles and he sticks by them even when they conflict. Frankly, I was just as conflicted.

The resolution came from the community. Snickers was adopted by a little girl, named Mary, of course. She fell in love with Snickers at the very moment her grandmother was looking for a project that would teach her granddaughter responsibility. Snickers is now a very happy lawn mower for a family in town. And yes, he follows Mary around.

And that's lambing. It's the joy of birth and the sense of renewal mixed with loss that is both blunt and mundane. One moment there is sweetness and hope and the next there is a cold callousness that demands we move on. And both dimensions have to be held in balance if you're to stay effective as a farmer.

While in between, there are these little stories that never travel down the path you expect, do they, Snickers?

HOT DAY TO HAY

In the coastal mountains of Oregon, most farmers turn pagan at some point in the summer. Early summer begins a polite beseeching of the rain god and the sun god to cooperate. Why can't we all be friends? The dialogue turns menacing by mid-summer. This is based on that delicate balance of sun and rain necessary to make hay. In the Northwest, heavy spring rains bring lush grass that is waist high by early June. The farmer surveys this abundance with that warm feeling of contentment that this will be a good year. He just needs the humidity to drop low enough to dry the grass after the cut. Wet grass molds in the fields and spontaneously combusts in the barn. Many a barn has been lost to wet hay.

Unfortunately, this is the coastal Northwest, famous for its cool mornings with heavy dew and mild afternoons. That's good for tourism but bad for haying. At a certain point, grass reaches a peak in nutritional value, and each day beyond that causes it to grow more rank. And each day that passes without a first cut pushes out the time frame for a second cut, eventually making a second cut impossible. Finally you throw in with the pagans and start looking for small animals or a tourist to offer in sacrifice. Something that won't be missed right away.

When the heat finally comes, those farmers with haying equipment are in high demand. As newcomers, we were last on

Will's list for haying. And haying is not a simple process. First the grass is cut and raked into furrows for baling. The furrows will be turned and fluffed over several days to speed the drying. Pray it doesn't rain. Once dry, the baler travels down the furrows, scooping, folding, and compressing the grass into bales that are cinched tight with twine and disgorged out the back. Balers are complicated pieces of machinery with lots of moving parts that are bounced over mole holes and rocks, which constantly throws them out of alignment.

This causes the farmer to stop his tractor and get down to make the necessary adjustments. Balers are well known for their insatiable appetite for fingers, so this chore is done with considerable care.

Once the bales are in the field, the pressure is on to get them out before it rains. Rain on the bales and the whole project is a bust—you can't dry hay bales. Getting the bales out of the field requires a team of sturdy young men to buck them onto hay trucks and then into the barn. The only good thing about this task is that teenage boys still want to look buff for teenage girls, and it's better to get paid bucking bales than to pay a gym to lift weights. Every year a buck team has to be assembled from the local pool of teen boys. Then we wait until the farmer has adequately appeased the gods of weather. And suddenly, one morning, it's go time.

It hadn't hit 90 degrees all summer, until the day we pulled our hay. Then the temperature soared, and the pagan gods laughed. We started at eleven, after the dew was lifted and the baler had a lead of several rows. I drove the truck and Greg lead the buck team. Both my daughters, home from school, were pressed into service with the admonition that "the family that farms together, eats together." By mid-afternoon, my eldest decided she would rather skip the eating, and if pressured anymore, skip the family too. Fortunately, a young girl from town had decided that the boys shouldn't have all the fun. She kept my youngest company in the sea of testosterone.

By five we had fifteen tons in the barn with seven tons to go. The boys were stripped to the waist and glowing red from exertion and sweat. This is where I had to be attentive to the signs of heat exhaustion, since Greg was in the same condition as the boys. We stopped regularly for rest and water. I was impressed with how the girls kept up, but the stack was approaching twelve feet high and only the boys, with their upper body strength, could toss the bales to the upper tiers. Still, the girls did their part of the work without quite realizing their bigger role in motivating the boys. The boys were on their game, as they subtly competed for the attention of the attractive young women.

It was during the swelter of late afternoon that I noticed the young woman looking flushed and dull-eyed, early signs of heat stroke. I called to the two boys beside her, "Is she hot?"

Both boys froze, turning a brighter red than I thought possible. I suddenly realized I was talking to teenage boys. "I mean, is she really hot?"

"Totally." This was delivered with rapt sincerity by the more gangly of the pair.

This exchange brought her back into focus and now she was flushing bright red.

"Oh, for goodness sake, get her to sit down and I'll bring the water."

My practicality seemed to break the spell. The boys looked sheepish and their peers began to laugh. My daughter shot me a look and rolled her eyes at my lack of teen awareness. I passed the water around and we got back to work.

The moon was rising when we finished. With each dollar I paid them, I thought of how it brought me closer to that hot bath and soft bed. As the boys walked off, counting their money, I could hear them making plans for the night. Then one of them, the gangly one who revealed himself with such sincerity, turned to the young woman. She was getting in her truck and he asked, if she didn't have plans, perhaps she would like to join them.

Summer nights are short in Oregon and we try not to waste them.

BARN PARTY

Even before the barn was complete, Greg knew it needed a christening. In the minds of men, something about building resembles birthing. And that calls for a recognition, a barn party. Greg claims to be the source of the inspiration, but I had no doubt that Jack supplied both the kindling and the matches. Jack was a folk artist, and his favorite medium for the crafting of folk was the party. For Luther, parties with alcohol surmounted his general objections to social obligations that might impose restrictions on his freedom.

So when our dark, dingy barn had been transformed into a brightly illuminated palace of solid timber construction, and when it was festooned with the sweet smell of fresh hay, the time of christening was upon us. The call went out. On a full moon night we would host a barn party. Everyone was welcome.

Of course, as with all things rural, we had no idea of what we were doing. How many people would come? How much food and beverage should be ordered? And the preparation! The barn would need to be cleaned, which is no small feat. Barns with haylofts means strands of hay come filtering down from the ceiling whenever anyone walks on the boards. Basically the barn looks like the underside of a hayfield. And we're going to serve food in there? I began waking up in the middle of the night again.

Jack stopped by to help me sort it out. The food was easy: a keg of beer, a tub of ice, a dozen sodas, a dozen hamburgers, a dozen buns, a grill, a tub of potato salad, a tin of brownies. Done. I noted that didn't sound like much food. Jack winked, and said, "These are country people. They grow their own food; it's people that's in short supply. But you do need to clean the barn."

The day of the party we pressed the girls into service again. My eldest, quoting her econ professor, noted the only reason farmers had children was for the free labor. She credited us with being real farmers on that score. Greg power-washed the barn. The girls strung Christmas lights. I ran to town for supplies.

Toward evening, Jack's truck rolled up. He unloaded sawhorses over which he put old doors, then tablecloths, and then a series of tented canopies. He looked at his banquet hall and judged it good. Next he unloaded an oil drum bar-b-que, stoked it with firewood, and set it ablaze. Then he took out a lawn chair, a cooler, and a milk crate. He positioned these items in front of the entrance to the barn, until they were just so. Finally, he went to the cab of his truck and pulled out a fiddle case, which he slid under the lawn chair. He plopped in the lawn chair, took out his pipe, then pulled a beer from the cooler, popped it open, and set it on the crate. He leaned back and lit his pipe, took a sip of beer, and smiled. He winked at my girls and gave his unique salutation, "Bright moments to you!" Both girls reported knowing from the moment he smiled, the party had started. His smile said it all. Something interesting was about to happen.

Which is probably why they came back to the house. It's a short stroll from interesting to dicey. I loaded them with a platter of hamburgers and a bowl of potato salad. I brought the buns and brownies and we arranged them in the center of the long banquet table. Surrounded by so much empty space, the food looked forlorn. *Foreshadowing?* I wondered. What if no one comes? We didn't actually invite anyone except Brick, Jack, and Luther. No one knows us, why should they come, even if they knew we were having a party, which they might not? I decide to distract myself by gathering raspberries for the brownies. At least there are brownies.

Except I found the raspberries stripped. Chipmunks, no doubt. Bezel, the cat, had given up on policing the outside terrain. So many variations of outside rodents, some quite belligerent. He confined himself to interior rodents, and mostly only those that could be surveyed from the sofa.

Brick arrived with his wife, Lucille, and their two children. They contributed a platter of elk steaks, a large salad, and apple pie. His children brought a bag toss game and began setting it up. Luther arrived accompanied by his dog and a watermelon carved into a bowl with three kinds (orange, green, and red) of melon balls inside. Sort of Martha Stewart, if she were in drag and dressed in overalls. Jack's daughter, daughter's husband,

and granddaughter appeared next. The granddaughter dressed as a fairy princess complete with wand. Then Will and his family and Nate and his family and Sarah and her family and . . . people just kept arriving with kids and dogs and handshakes and hugs . . . and the banquet table was full to bursting and the grill was filled with every kind of sizzling meat . . . and the aroma. Oh lordy, we had a party going on.

The men set up horseshoes, and the children built forts out of hay bales in the barn loft while the women circled chairs around the campfire. And between the various camps ran a bevy of dogs on the prowl for unclaimed plates.

As the moon climbed in the sky and the plates of food dwindled and Luther's 100 proof melon balls took effect, Jack slid his fiddle from its case. He drew the bow sharply across the strings and announced, "Let's see about this," and launched into a jaunty bluegrass reel. Several guitars and a mandolin stepped up and the music surged forward in a rolling stomp. People were on their feet, clapping and dancing. From their stalls, the horses appeared to be bobbing their heads to the music. I noticed Will sharing his beer with Cisco, the bad dog, in a moment of deep, fraternal understanding. I saw Lucille curl up in Brick's lap while they shared a bite of lemon meringue pie. And I watched teens slipping out from beneath the bright lights and adult gazes to explore the elusive mysteries residing in the shadows down by the creek.

At about this time, my husband encircled me from behind in his arms, and nuzzling against my face, asked with melon-ball breath, "Are you happy tonight?"

I allowed that I was and we swayed a bit to the music. And I thought about my life in Phoenix and how we would never

Leaping Lamb Farm from hayfield

Leaping Lamb Farm 1895 farm house

The main barn, built in 1932

Chaco and Mora

"Hair sheep" rams

Annie's house

Bridge over the creek

Loading hay into the barn

Chaco

Scottie and "Rabbit"

Annie and Caitlin with Chaco

Greg holding twin lambs

Processing apple cider on a cold day

Our assembly line setup for inoculating mushroom logs

Petting a chicken
Photo credit: Joanna Lezak

Peeping Tom at the back door

Henry (farmer in training) delivering produce from the greenhouse

Tater

Paco

Patches (the good dog) and Cisco (the bad dog)

Turkey poult
Photo credit: Shawn Linehan

Harold the day after the attack

Deedee and Nona at two days old while they were still living in the house

Shearing in the barn

Tater scattering the sheep

Photo credit: Dennis Rivera

Photo credit: Paul Deatherage

Scottie and Paco
Photo credit: Alexandra Shyskina

Herding sheep
Photo credit: Shawn Linehan

have had a party like this. In Phoenix, the party would have been stratified by age, class, and lifestyle. Here, you don't host a party as much as just offer a platform. Everyone comes and offers a little something, making it a cooperative event. And I liked that. I liked feeling a part of something.

"Whoa there, we like to see two fingers of separation between couples if you don't want to get called out in church tomorrow."

I turned and saw Jack munching on a brownie. He had surrendered his fiddle to younger talent. My inebriated husband reached over and broke off a piece of brownie from Jack's ample portion. Jack responded with feigned umbrage, "Well aren't you the grabby Gus. I don't like to think where those paws have been."

Greg started to respond to Jack, but redirected at me, "I thought you were going to put raspberries in these?"

"Raspberries!" Jack exhaled. "I had hoped for so much more." He handed the rest of his brownie to Greg.

"There aren't any raspberries, or blueberries, or any berries. What there are are well-fed chipmunks. We've got to come up with a better plan."

"Why don't you get a cat?" Jack offered casually.

"We have a cat."

"No, I mean an Elsie cat. Not that yuppie furball you got now. Most cats were domesticated a couple thousand years ago. Elsie cats, more like . . . well, hell, they aren't domesticated. That's the point. They're descended from cats city types bring out to the country to throw away. The ones that survived are Elsie cats."

"Come on, that sounds like a rural myth." Greg has a strong skeptical streak.

"Suit yourself. But I saw a card in the Merc advertising free kittens. Cats out here tend toward the feral—kind of like the people. And speaking of which, your youngest looks like she's taking to country life. I'm not sure about your oldest."

He was motioning toward my girls sitting on hay bales around the campfire talking to a couple of young farm boys. They were all laughing and appeared to be having a good time. I'm not sure how Jack picked up on it, but he was right. Annie showed a genuine interest in the farm while Caitlin couldn't wait to get back to her life in the city.

"I think that may be true."

"So when she's done with college, think she'll come back here?"

"I'm not sure. I don't know what job there is for her here."

"Same as everybody else. Live here and work in the town."

"I hadn't actually thought much about it. I've been so focused on whether we're going to make it." It was a good question and I looked at Greg.

"Yeah, same. I don't know. I can't quite see her living as an adult under the same roof as us. I think she would want her own space."

Jack lowered his plate as an idea began to rise up in him. "Sounds like you might need another house. The county allows you to add one more building for farm help. And I do know that the Wayfaring Brothers Construction Company has some availability in their spring schedule. I would have to coordinate with my partner, when he sobers up, but I think we could find a place for you in our schedule."

Greg looked at me. "It would increase the value of our property."

"Yeah, except we don't have the money and we don't know if we're going to survive here ourselves and every idea looks good after Luther's melon balls."

Jack looked up, "Speaking of which, did you notice how many balls were left?"

Apparently that was the cue as they ambled off in tandem to the melon bowl. My head was starting to swirl as I thought of Jack's proposal. What would happen as we grew older? You can retire from a job, but farms are something you want to pass on to your children. Who would take over? And if not our kids, why were we working so hard?

And then I saw people dancing and people sitting around the campfire laughing and I was reminded how I could not duplicate this experience in the city. And then I saw my inebriated dog about to pee on a hay bale with two of my neighbors perched on it. And I knew there were experiences I didn't want duplicated. I rushed to chase him off.

I sat down next to Sarah and sampled melon balls off her plate. And we talked and we laughed late into the night.

Greg was right. Our barn needed that party.

A HAROLDING EXPERIENCE

The back of our farm opens to the wild. Beyond our farm gate is a logging road that ends in a forest. Miles upon hundreds of miles of wilderness. It was only a matter of time before the wild crossed the gate and found its way onto our farm.

It was a lovely spring morning. We were just finishing our walk when Sarah noticed something crumpled and white in the brush. It was a lamb, damp with blood, unable to hold its head up, but alive. I scooped it up and trotted back to the barn where I could tend to it.

Its ear was nearly torn off. Shaving the wool off with clippers, I found puncture wounds around its neck. I washed the wounds and dressed them with blue lotion, an antiseptic that turned the lamb's head and neck a deep purple hue. Greg arrived and noted the broken front leg. The break was complete, allowing the leg to twist in any direction. The lamb made no protests as we inspected it. We splinted the leg with a stick, taking care to immobilize it well above the break, just as our sheep book said to do. Then we laid the lamb down in fresh straw with a heat lamp. The next twenty-four hours would tell.

Clearly, the lamb had been the target of a predator—most likely a solo coyote. We guessed that our morning chatter on the trail had diverted the predator from its goal. The lamb was too large and unwieldy for a rapid escape so it was deposited in the bushes. The question now, was it the only victim? I did a head count while Greg checked for tracks. It would be good to know what we were up against. Rarely do predators take only one helping. Once they know the buffet bar is open they return until there's a reason not to. It's the job of the farmer to supply that reason—other than an exhausted supply.

Counting sheep is difficult. They let me get up to the last four or five and then the whole herd does a stutter-step shuffle, half to the left and half to the right. I'm pretty sure it's on purpose. If sheep are difficult, lambs are impossible. They meander under their mothers, seeming to disappear in the woolly understory. Often they've ducked down for a little refreshment, so it's not counting heads so much as counting butts. And if you've seen one butt . . . well, you know they don't offer much individuality. Even allowing for the disappearing lamb discrepancy, the count was low by at least five. Our little convalescent was not the first victim.

The news from Greg was equally dismal. He had found cougar tracks. Far more efficient predators than coyotes, cougars are the bane of sheep farmers in our region. And they enjoy a protected status so they can't be freely hunted. A cougar in the act of taking livestock can be shot, but it's extremely rare to catch them in the act. The state offers the services of a professional trapper, but usually the trail is cold by the time the trapper arrives at the scene of the crime. Cougars range about 150 miles, and after a kill they move on, so the cougar is usually long gone by the time the trapper arrives. Of course, they will return. The buffet is open.

Still, it seemed strange that a large cougar had dropped the small lamb instead of carrying it off. The answer came from a check of our birthing records. The lamb's bigger brother was among the missing. Apparently, hearing our approach, the cougar could not manage the twin pack he was carrying. Making this little lamb incredibly lucky—assuming it lived through the night. Glancing at it, that didn't seem very likely. Once again, nature delivers a losing proposition all the way around—for the farmer, for the cougar, and especially for one little lamb.

Before going to bed, I decided to check on the lamb. I prepared myself for the likely outcome but tucked a warm bottle of milk under my arm just in case. Truth be told, I don't like the feeling of surrender. I'm a little stubborn that way. The lamb lay prone on the straw but was still breathing. I trickled a little milk down my finger and into its mouth. The lips moved as it lapped at the milk. So I trickled a little more and the lamb lapped a little more. I gave the lamb the entire nipple and she began to suck. I smiled. This one was going to make a fight of it. She only got a third of a bottle down before she turned her head away exhausted, but it was something—a good thing. If this lamb could survive the jaws of a lion maybe I could survive the doubts and down-drafts that threatened to swallow me. Maybe.

We spent the next day hiking the property in search of stragglers. Now that was a vain hope! Rarely do lambs separate from the herd. Occasionally lambs get caught in blackberries and remain caught until we hear them bawling and cut them out. With a cougar on the prowl, we would not be the first responders to a bawling lamb.

The following day we saddled horses to extend our search. The dogs came along. We lost Cisco in a particularly dense thicket. It took quite a bit of calling before he came out, but eventually he did. That was not unusual for Cisco. He is the bad dog.

Brick heard of our troubles and was eager to help. He brought his dog for a second canvassing of our property. This time all the dogs vanished in the same thicket. Brick bushwhacked his way in to inspect and I followed. Halfway into the thorny maze, he found lamb carcasses. Cougars don't eat their prey at the kill site. They drag it back to a lair where they can eat in peace and return for seconds. This was the perfect place to hide cougar leftovers, so dense it was nearly impossible to move in any direction but one.

It suddenly occurred to me that we were in the cougar's pantry. And we were unarmed. I suggested that we not overstay. Brick was charged up on the idea of hunting cougar until he realized at that moment he was more likely to be the hunted. We agreed that lingering served no purpose.

The state trapper was out the next day with his dogs but, as expected, the trail was cold. The cougar was long gone. The trapper cautioned us to be vigilant over the next two to three weeks as the cougar completed its foraging loop. Most likely a young cougar, inexperienced at hunting, and fallen into the easy path of harvesting livestock. And like most things that come easy, the long-term consequences were going to be more severe—probably fatal in this case. Close association with humans almost always ends badly for cougars.

As the days progressed, so did the lamb. After a few feedings, she was on her feet hobbling around. I moved her to the kitchen infirmary to make feedings easier for me. Her peg leg on the wooden floor combined with her purple and white, two-toned complexion kept the dogs in a nervous state. The mere sight of the zombie sheep approaching them sent them flying through the dog door.

I named her Harold in homage to the children's book *Harold and the Purple Crayon*. To me, it seemed as though she had used a purple crayon to recreate herself. The name captured all the incongruities this plucky little lamb encompassed. She became a favorite of mine, always reminding me that if she can survive the jaws of a lion, then what can I survive? She would live a full life on our farm.

We continued to lose lambs to the cougar that summer. All our attempts to redirect, trap, or kill failed. We strung an extra loop of barbed wire to raise the height of our fences, but the cougar still leapt over, carrying off hundred-pound lambs. We erected a radio to blare rap music and kept lights on the loafing shed all night. We cleared thickets to deny sanctuary. Greg took to sleeping in the woods with his rifle. All to no avail.

That September the killing stopped. We heard a cougar had been shot in the valley south of us. But we heard of cougars being shot every summer. Perhaps it was our cougar that had been shot. Perhaps it just moved on. We'll never know.

The back of our farm opens to the wild. It's only a matter of time until the wild finds its way onto our farm. Again.

PART FOUR

NEW YEAR WISER

This was our third year tally and I came to it chastened by experience. I no longer expected the farm to be profitable, but I did hope for a trend in that direction. A little better than last year would be good enough. Unfortunately, the price of heritage turkeys was down. Other farmers had discovered the niche. The price of lamb was up, but the switch to Katahdin had lowered my harvest weight. It was going to take several years to build the Katahdin brand sufficiently to command a price premium. Then there was the cougar, which had lowered my total count.

Total farm income was down significantly from the previous year. But the deeper trend was promising. We now had two niches that could eventually make the farm sustainable—but how long would that take? We were still burning through our retirement savings. Greg quipped there was no need to worry since, at this rate, we would die before we retired. I didn't laugh.

The unholy trinity of farming states, "It is a high-capitalization, low-profit, high-risk business." I doubt there is any other business that requires as much up-front capital to operate while returning so little profit and bets it all on something as capricious as mother nature. Big ag has the scale to exploit the razor-thin profit margins and, with access to government subsidies, to manage the risk. Small ag cobbles together broken-down equipment to scrimp on capital, chases niche markets for profit, hopes for mother nature's bounty, and scrapes by in the meantime. Play for time and hope for the best.

Of course, hope is not a plan, it's a wish. Too much wishful thinking leads to delusions, which, in turn, lead to corrective consequences. We had shed a lot of our delusions since coming to the farm, but maybe the farm itself was the final delusion to be pruned.

This year's numbers argued for capitulation, but the deeper trends offered a taunting "maybe." After all, the numbers were based on this year's market, and markets fluctuate. I would regret quitting the farm if next year's market turned favorable. Probably the deeper truth was that I wasn't done with the farm. Something inside told me I could make this work. I just needed time to give the trends a chance to manifest. I needed time to grow these niche markets. Quitting before that would leave me ruminating on "might have been."

Was this perseverance through adversity or stubborn adherence to a delusion? I couldn't say. Certainly, farmers are known for their stubbornness, possibly because the farm life continually confronts them with decisions like this one. The numbers are mixed, and the data is ambiguous, forcing the farmer to look inside and subjectively decide an outcome. Those who look inside and say, "enough," leave farming. Those who stay become stubborn old farmers.

Was I becoming stubborn? Probably. And tough too. I was being shaped by repeated exposures to loss. Every year I lost lambs and every year I felt sadness with each of those losses. But increasingly, I was looking to the next year's season with a growing sense of the ebb and flow of life that assuages that sadness. The loss of a solitary lamb was no longer a singular event, rather I viewed it in the larger context of nature's cycles. I didn't know if I was becoming more resilient or more calloused, but certainly the experience contributed to an attitude of defiance against adversity—a willingness to play for time, work to improve the odds, and wait for a better outcome. I could take loss, even multiple losses, and bounce back.

One of those deeper trends that kept me hopeful was our connections in the community, which included a roster of skilled tradespeople to call when the need arose. Just as important, we were known to them, which helped with communications. Chief among these were the Wayfaring Brothers. I really needed to find something for those boys to do that would make money for all of us. There was Jack's suggestion of a farmhouse for my daughter.

Farm and family go together. It's hard to justify the long hours and hard work if it's not part of a legacy we build for our children. Probably most business owners would like to see their children take over their business. But if the children are not so inclined, the business can be sold, and the family home remains. Farming is different. It is business, home, and lifestyle all intertwined, and legacy is the glue binding it together.

This year, our eldest daughter had graduated college and taken a job as an administrative assistant in Seattle. A job that did not require a college degree. When we inquired about the rationale, she reported she wanted to take time "to enjoy life." It wasn't the job, but the Seattle lifestyle for young singles that called her. Apparently, offering our lives of hard work and frugality as examples did not persuade her that this was the path to happiness.

Our youngest daughter, Annie, was in her final year of college, majoring in animal husbandry. Her choice of a major was her choice alone. We remained stoic in her presence so that she could be clear the choice was made without influence. She demurred about her future, saying only that she was focused on finishing college before deciding a direction. Of course, secretly we were proud and hopeful about our joint future. I could only imagine the improvements to our herd she could make. We just had to play for time.

So here we were in our third year, unable to show the farm as a sustainable entity. Yet if it was to be sustainable, we must build for the future, no matter the debt. By the end of the evening of accounting, we both knew we had to build a house for Annie. Even though she was not decided on her future, we must build it to create the possibility of legacy.

Well, why not? We had already leapt into the unknown with the farm. If you're going to leap, leap with faith.

So, Wayfaring Brothers, you've got a job. But it's not a house, only a small cabin to safe-keep a possibility.

MICE IN MODERATION

Old farmhouses come with a resident mouse population. After all, the mice were there before the house and have simply adapted to the structure that was erected over them. Compared to an open field, the farmer's pantry is a mouse jackpot. For the farmer, the jackpot looks more like felonious theft and vandalism, which requires countermeasures. Most farmhouses are mined with traps. Given the abundance of mice and the high value of the pantry, the best the farmer can hope for is a "mice in moderation" policy. I have come to accept mice in moderation.

This delicate truce is upset every winter when the house mice are dispossessed by their country relations. The winter population swell pushes the mice to brazen acts of delinquency. This is not to be tolerated. Even Bezel, the grumpy cat who had retired from hunting in favor of the perpetually filled food bowl, has, on occasion, been provoked into killing mice that disturb his slumber.

As for me, I doubled down on traps. Not just the pantry but every cupboard and closet, including one cynically concealed beside the cat's bowl. I was still hoping the cat might see the irony and reclaim his self-respect. But the only reclaiming Bezel did was his spot on the sofa.

Typically, this intensified assault snares several mice the first night followed with declining results thereafter. By the third night, half the traps were sprung with nary a mouse carcass in sight.

By the fourth night, some of the traps had gone missing. It seems a physical impossibility that a mouse could drag a trap through a hole . . . unless it's a really big hole and a really big . . . well . . . Rather than think about it, I decided to anchor my traps with strings and tacks. That stopped the pilfering of traps, but mouse mortality remained unacceptably low.

It was time to change it up. At the Merc, Nate offered a shiny new trap with a fake cheese plate that didn't require baiting. Since my jar of mouse peanut butter was getting low, this might be just the thing. Sure enough, I caught a mouse the first night, and the second night too. The third night my streak went cold. I changed locations. I left enticing crumbs. I cleared paths for easy mouse access. Nothing. Perhaps I had trapped them all? Except there was mice poop in the mixing bowls and the cat food was disappearing at an alarming rate. The mice were laughing at me.

I needed a nuclear option. Maybe Nate had something under the counter. Something not quite legal. If not, I bet he knew someone who did.

I never made it into the Merc. I was stopped at the entrance by a cardboard box labeled, FREE TO GOOD HOME. Inside was a green-eyed black cat. I flashed back on Jack's advice, "what you need is an Elsie cat." Could there really be a difference? Did I really have a choice? I reached in to pet it, and it began to purr. I thought about Bezel and began to worry he might feel displaced. Then I thought about the mice. It's a farm. Everyone pulls their weight. Even grumpy cats.

On reflection, I probably should have paid more attention to the sign. While it said FREE TO GOOD HOME, there wasn't anyone there to ensure compliance, suggesting "good home" was broadly defined.

And then there was the introduction to the family. That too, was probably a foreshadowing of things to come. When Greg offered a tentative, get-acquainted stroke the cat purred and brushed against his leg. This act of affection was reciprocated by Greg, who offered a more committed stroking of the cat's back. Apparently, this was the act of complacency the cat had been waiting for. It snapped back, both claws clutching its target while needle-sharp fangs impaled Greg's hand. Greg was not pleased. The cat released its victim and just stood, without contrition, calmly surveying the room.

Our resident cat, Bezel, took one look at the intruder and went into a three-octave hissy fit. The new cat responded with a predatory stare, leaving Bezel to slink up the stairs. Patches came over for a sniff and got raked across the face, sending her out the dog door. Cisco began a low menacing growl while avoiding eye contact altogether. The cat walked slowly around Cisco, found Bezel's bowl, and helped himself.

I felt like I had just witnessed the feline version of a Clint Eastwood movie. The dark stranger had put everyone on notice. Greg was giving an emphatic, and swollen, thumbs down to the new addition. "Okay, I understand your reluctance, but we've got to do something about the mice. Let's give it a night or two."

The next morning there were three mouse carcasses, in various stages of dismemberment, scattered through the house. This was promising, but I remembered the pattern was always good the first night. The second night was punctuated by cat paws bounding across wood floors, those little mouse squeaks, and the thump of mouse bodies hurled in the air and hitting the floor. Then, silence. It was another three-mouse night. Every night thereafter was a three-mouse night. I had an Elsie cat.

Greg took to calling the new cat Bubba, because he seemed satanically inspired. Since we already had a Bezel, might as well complete it with Bubba. Two ends of the satanic spectrum—sloth and slayer.

After the first few weeks, Bubba had depleted the inside mice population, so he took to thinning the outside. A variety of other dismembered rodents began to appear in the house. Disgusting, but I rationalized, better than having mouse poop in your granola.

Then I found mouse poop in my mixing bowl. Not possible. Probably a leftover from the Bezel era. I found more poop in more places, and the cold truth had to be acknowledged. There was a mouse in the house. A little detective work revealed the source: Bubba. Like any good sportsman, he had adopted a catch-and-release policy. He was bringing mice into the house and releasing them, mostly for his own late-night amusement.

So now you understand. It's that eternal compromise with mother nature. Even with an Elsie cat, the best you can hope for is mice in moderation.

A PLACE FOR ANNIE

We decided on a place for Annie. Something that would call her home—that is, call her to a home she never knew as home. We hoped it would call her to a possible home.

Daunting as that task might be, Oregon land use laws made it harder. Oregon was founded by farmers who trekked the continent to be able to farm here, and ag culture continues to have a powerful voice in Oregon politics. Oregon's farmers were among the first to recognize the dangers posed to farmland by urban sprawl. Real estate developers have the dual effect of increasing the price of land and the property taxes, in effect pricing farmers off their land and turning valuable farmland into lawns. Remember, the efficiency of farmers has made farming a thin-profit enterprise, and that means land has to be cheap for farmers to compete. To protect farmers, Oregon passed some of the most restrictive land use laws in the nation. Essentially, no new homes can be built on a property unless it is within an "urban growth boundary" designated by an Oregon city.

Farms are allowed a limited exemption to build a house for farm labor. And that was exactly our intent. It takes as much legal work as wood work to build a house on a farm, but we were approved. Now came the hammers and saws.

Initially we thought to purchase lumber and have the Wayfaring Brothers build the cabin, but Brick interceded. He reminded us that we live in a lumber yard. The only difference between a tree and a house is a saw and a man who knows how to use it. Logger logic. What seemed a large-scale industrial operation to us was, for Brick, a weekend do-it-yourself project. No need to fell trees when the wind has done that for you. Just hook a choker cable to it and haul the log to the mill. Easy peavey (a *peavey* being a wicked-looking tool with a long pole and iron claws for moving logs).

Okay, but a log is not a board. And that's why loggers have friends who are millers. Mobile Mike had a mill on wheels. No need to haul logs to the mill when the mill can come to the logs. So, with Brick's help, Greg bucked a dozen trees and dragged them off the hill. Mike set his mill up in a clearing beside the creek.

For five days the mill sawed logs into boards. Greg used the tractor to push the logs onto the mill's giant claws. The pneumatic claw lifted the log onto the mill's track where metal teeth held it in place as a band saw traveled the length of the log, stripping off boards. Mike sat in the cockpit monitoring the speed and tension on the blade. Too fast or slow caused the blade to bend, scalloping the board or breaking the blade. Each log presented a mathematical puzzle to determine how many boards could be harvested. Dents and bows in the log had to be compensated for. And each board had to be pulled and stacked with spacers placed for drying. Jack was brought in to help with the stacking. Or storytelling, depending on your point of view.

As with most things ag, the days were long, and the work was both hard and tedious. Cedar was cut for planks and sheathing, fir for studs and beams. The men were covered in sawdust and sweat. Mike loved his job and insisted everyone take a turn at running the mill, drawing the men into shared work. Around this, Jack wove stories that invited others' stories.

This is how homes are built in Elsie. With hammer and saw and diesel and wood and with great loops of stories that hold them together. The sore backs, mashed fingers, and wood splinters offset with horseplay and jokes. All for the house that would call Annie home—to a possible home.

STEPPING UP BY SETTING DOWN

The truth is always there—silent, resolute, immutable. There it stands, waiting for us to acknowledge it. Yet we defer. Perhaps it's not the truth we resist so much as what that truth will require of us. Better if we can slip past to the other side where we imagine the grass is greener. And of course, we can do that. We can always step around an uncomfortable truth.

So when Greg noticed my horse Chaco stumble, I noticed how infrequently it occurred. When Greg saw Chaco clip a corner or bang his head, I saw how he had been distracted or bumped by the other horse. Had I looked, I would have noticed how he always followed Mora out of the corrals. I would have noticed how Chaco, a great jumper in the arena, never jumped on trails.

I waited until the truth stood in front of me one evening. I was about to ring the bell that calls the livestock in from the fields when I noticed Chaco was already there. Very odd that he would have separated from his partner. Then I saw the blood dripping from the gash in his chest. He was also missing a shoe. After doctoring his wound and locking him in a stall, I went in search of a cause. It didn't take long to find the broken fence post and snarled barbed wire. This fence section was near the tree line and in deep shadow. He apparently walked into it and then panicked. The truth was that Chaco was going blind.

In all probability, his sight had been diminishing long before we moved to Oregon. It just wasn't apparent in the predictable world of riding arenas. In Oregon, the deep shade of trees merging with the dark green moss made the terrain void of form. Chaco developed a strong preference for level pastures and the close proximity of his trusted navigator, Mora. And the sheep knew to stay out from underfoot.

Until now, that had worked reasonably well. I inspected his wound, which was relatively superficial, and decided he could still manage his world. I would leave his chin whiskers long. At least he could feel his way around the paddock and he showed no problems finding the hay in his bin.

Greg saw the same truth and drew a different course of action. How does a horse survive when he can no longer see? How many fence entanglements to escape, painful gashes endured, or bone-breaking falls lay ahead? He already refuses the bridge because he can't see the edge. Would he be swept from his feet that winter by a raging creek because he wouldn't cross? What does our love count for if we can't be responsible for relieving the suffering of those we hold dear?

Chaco was my horse. I had spent many hours working with him. Those hours forge a bond between horse and rider—an instinctive sensing of each other's body that results in a blended will producing a single fluid action. The rider signals a direction but the horse chooses the implementation. A good rider yields to the horse's action as much as the horse yields to the rider's direction. It is a cross-species relationship that is both unique and powerful. That is to say, it is a relationship that does not brook interference—not even by a spouse.

Apparently, I effectively conveyed that sentiment when I suggested to Greg that he take his negativity and peddle it elsewhere. He sucked in his breath and walked off. Chaco was fine. He wasn't completely blind. He could see shapes and forms. His partner, Mora, steered him onto the pasture every day. She was getting older and wasn't taking any wild runs through the woods either. They were a team, both compromised by age and both compensating for the other.

As I thought about it, they weren't the only aging team on this farm, managing to get by through mutual compensation. Who can say when it's another's time to die or how much pain makes life unbearable? Or how that equation changes when those charged with caring for us actually care for us—in heart and deed? So not to worry, Chaco, I will look out for you. And Farmer Jones, you might think about how you'd feel if your caretaker adopted your policies? When do your disabilities outweigh your value? When should we put you down? Consider that.

The call came in the middle of the night, as those calls always do. It was my sister. Our parents had been in a terrible auto accident. Dad was in the ICU.

My flight back to Connecticut was divided between nostalgic memories of my father and my fear of what I might face in the hospital. My father was that rare man more comfortable with nurturing than directing. It was my father who made breakfast every morning, tucked me in at night, and assuaged every bad grade with "don't worry, honey; you'll do better next time." It made him very popular with my friends, who marveled at the contrast with their rule-enforcing fathers. So while my friends feared upsetting their fathers, I worried about disappointing mine.

I think that made it harder when I saw that gentle man lying in a hospital gown, unconscious and hooked to life support. He was so frail and so dependent on us, his family, for his care. Mom was in another section of the hospital recouping from lacerations and broken ribs. She was barely conscious herself and heavily sedated. So it fell to us, the children, to be adults.

My father was a fastidious man, known for his attention to the details in life. He was the first to have his sidewalks swept after a snowstorm as an assertion of civic pride. The first to jump up after a meal to wash the dishes as an act of family fidelity. And he was a real stickler for proper grammar. We all suspected he slept with his pajamas buttoned to the collar—possibly adorned with a bow tie. No one knew for sure because no one ever saw him in his pajamas. He came dressed to breakfast.

So I was shocked to learn from his lawyer that such a fastidious man had not prepared an advanced medical directive. There was nothing to assist us with his wishes at this difficult time. Perhaps for him, this was one of those truths that was easier to step around.

By the second week his condition was deteriorating. I was having trouble aligning my memories of the man who once held me in his lap with the bundle of broken twigs under that hospital gown. The surgeon and the social worker were making it clear that we, the family, would have to make that decision for him. The hope for recovery had passed, and the machines were artificially prolonging his life. What followed was a very uncomfortable discussion. Which child has the authority to make that decision and based on what criterion? You can hope for unanimity, but with four of us, total consensus was more ideal than real.

After a long day of discussion, we came to an agreement. We would meet the following morning to turn off the machines. When morning came, my sister had reconsidered. What followed were intense feelings of frustration compounded by ancient memories of childhood betrayals and slights that threatened to split the family irrevocably. We elected to step away and discuss it further the following day. I secretly feared the next day could be worse. That night, my father, the soft-hearted pacifist, slipped away, easing the family tension just as he always had done.

We buried him with a fitting celebration of his life and legacy. Mom was still mending. It would take her months to overcome the shock of this catastrophe. She would never be free of the grief.

I flew home more divided in my thoughts than when I left. I had always secretly thought that my father's supreme gentleness was a cloak protecting him from the ugliness in life. Who would want to harm such a tender soul who gave no offense? But the cloak offered no protection whatsoever. His death was violent, premature, unnecessary, and unfair.

Greg tried, in his way, to support me during this difficult time, but he lacked the sensitivity of my father. And that difference only made me miss my father more. As Greg lay sleeping with head cocked against the plane's window, I was reminded of how he had pressured me about Chaco—something my father would never have done. And I resolved that Chaco would live a full life under my care.

Chaco was fine. He had stopped going over the bridge when he had misjudged the edge and fallen into the creek. Now he was refusing to ford the creek. The flickering light reflecting from the water made him unsure of his footing, so he stood by the bridge, alone, waiting for Mora to return in the evening. There was graze on this side of the creek and he was making his adjustments.

Eventually he refused to leave the paddock. He would search a spot in the sun and just stand there. Alone. All day. At night, we kept a light on in his stall so he wouldn't hurt himself.

Jack had observed Chaco and offered his assistance. Knowing Chaco was my horse, he was tactful enough to suggest it to Greg. He knew horses were more pets than livestock and for that reason it was easier for a third party to offer a merciful end. Jack possessed a pistol and he knew where he could borrow a backhoe. The whole thing could be over in an hour. He left his information with Greg and Greg had the unhappy task of conveying it to me. I saw it for what it was. Two men trying to bend the world to their singular, utilitarian needs. Not on my watch. I would continue to keep a place for caring in the world. And I cared for Chaco.

The call came again in the dark of night, rousing me from deep sleep. Except, there was no call. Had I dreamt the ringing phone? I sat up and listened. There was only silence interrupted by my husband's ragged breathing. Then I heard it, a slight flutter of wings. The cat leapt off the bed and began to stalk in the direction of the noise. In a dark corner I caught the outline of a bird. It had been a warm night and we had left a window open.

I rose to investigate and found a small owl with a round face and yellow eyes. To my utter amazement, it allowed me to cup it in my hands. It remained quiet in my grasp, staring at me with a beautiful, calm look. I carried it to the window and released it. It took flight on silent wings, disappearing into the night.

Back in bed, I rationalized it must have been a juvenile, too inexperienced to be afraid. Maybe. But another part of me was convinced that I had just been visited by my father. There was no protest or conflict with this owl, only a calm acceptance. I had been carrying the feeling that I had let my father down in those last days. There was something more that should have been done. Irrational, yes, but feelings have their own reality. The owl reminded me of my father's unwavering acceptance of me. He was accepting in all things, but especially of me, and this bird's visit carried that message.

I began to see how unaccepting I had become. My guilt over being helpless had made me unable to see my father's need for liberation from a terrible encasement in a painful body. The visit released me from my guilt, replacing it with a serene feeling of acceptance. I was able to see that perhaps setting down my sense of control allowed for nature to take over. I was able to forgive myself for not taking better care of my father—for letting him die. It never was in my power to make him live, no matter how much I loved him.

It took a few more days but I could feel the ice around my heart breaking up and melting. There was Chaco, alone in the paddock. There was Greg doing chores, trying not to notice Chaco alone in the paddock. And standing there all along was the truth, waiting for me to accept it.

I went down to where Greg was working and asked him to call Jack. He looked at me to take the full measure of my meaning and then just nodded. I was thankful he didn't belabor it—but Greg was sensitive that way. I had been too busy stepping around that truth as well. Those who love us often get splashed with the emotions we're using, at that moment, to paint the world. I had made Greg insensitive so I could avoid the loss of Chaco.

For Chaco, it was another normal day. He ate grass. He was brushed and fed treats. His mane and tail were combed. Then we walked him across the creek to the far field where Jack had dug a hole. I said my goodbye and headed back to the barn. It is a hard thing to take a life, and Greg felt it was a duty that should not be foisted onto Jack. But Jack insisted, knowing it was harder for Greg. So Greg agreed to hold Chaco's halter.

Jack squeezed the trigger but nothing happened. He said it had jammed, so Greg was sent to the house to fetch his pistol. As soon as Greg was out of sight, Jack released the safety.

The shot echoed, spooking Mora. She took off, running, the sound of hoofbeats trailing the echo into the distance. The sheep lifted their heads in unison and, not seeing a threat, returned to grazing. One moment there is a beating heart and the next, there is silence. The tractor's diesel engine turned over. The hole was filled and the field returned to its bucolic, green roll. If not for the smell of fresh dirt, I could have closed my eyes and thought I imagined it all.

A week later, Greg asked me to walk with him into the woods. Deep in the woods is a magnificent maple that dominates the landscape. We call it the Grandfather Tree because its longevity calls us to contemplate our nameless ancestors. Hanging from one of its branches was a small figure of a horse woven from cedar bark. A gift left by Laura.

On another branch, hung an owl. I was dumb-founded. It was not possible for Laura to know of my visit from the owl. Chaco she could have witnessed, but not the owl. Greg nodded in understanding. "I found the horse several days ago and told Jack. I think he made the owl. Laura's not the only bark weaver in these woods."

So it's not all silence and emptiness. There is a grandfather tree bearing witness to all things past and present. And from its branches, memories of those I still love dance in the wind.

That night, by my window, I listened to the croak of frogs and the scratch of crickets. I was reminded that with friends and family, life is made rich. I strained my ears to hear the distant hoot of an owl . . . but what I heard, when I closed my eyes, were all those beating hearts. I heard them all.

OF TATERS AND KINGS

The heavy rumbling of a diesel engine brought me running from the far field, but I arrived too late. A woman, near my age, was jumping back into her truck. I began waving my arms to call her back. She gunned the engine and sped away. Odd, I thought, that she didn't stay long enough to identify herself. Odder still, when I turned and saw a large bay tethered to the hitching post of our barn. Someone had delivered a horse to the wrong address.

He was something. Sixteen hands high, with big shoulders and bigger rump. The kind of power quarter horse that cowboys have been known to sacrifice marriages, and even pick-up trucks, to own. He had a kind eye and social disposition. In fact, from first look, I couldn't find much wrong with him. Someone was going to be upset when he didn't show up wherever he was supposed to be.

Greg popped out from behind the barn, a little late as usual, "What's this?"

"Someone needs a new GPS. Wrong delivery, I guess."

Greg held him by his halter and inspected him, "Don't suppose we can claim 'finder's keepers,' huh?"

"Wouldn't that be nice? He's a good-looking horse."

Greg turned to me and asked directly, "You like him?"

"Yeah . . . wait what?"

Greg had that smug little half-cocked smile he smiles whenever he is feeling clever and self-satisfied. The smile was announcing the true author of the botched delivery. I decided to slap that smile off his face with a grateful kiss, before returning to inspect the horse.

"He's got to be expensive. We can't afford him."

"Probably not as expensive as you think. He's a bit green and needs work. And the sellers were motivated, so I got a good price. Saddle up. We've got 'til tomorrow to take him back."

"What's his name?"

"Tater."

"What?"

"Well, Tater-Tots, which rhymes with Potts, which is the family's name."

I looked at this magnificent horse with the big, brown eyes. Never was there a greater mismatch between matter and moniker.

"Well, we'll have to change that." I gave him a pat on the neck and went for my saddle.

He proved responsive to my commands and we settled into a lovely ride—for the first hundred yards. Then he stopped, dropped, and rolled. No warning, just an elevator drop to the basement. I got off on the first floor. Obviously, this little trick had kept riders off his back in the past. I registered my disapproval and we continued with the ride. I began to wonder about the sellers' "motivation" and the rapid exit they made. Almost as if they couldn't get away fast enough. But who names a horse Tater? Novices! Just the kind of people who let horses acquire bad habits. Well, a little work and I'd have him slicked out, smooth as frog fur.

Tater took the next week to get adjusted to the farm. First was the introduction to Mora, our mare. After the requisite sniffing, nickering, squealing, biting, bucking, racing, and neck jousting, Tater established his dominance, and life in the paddock settled down. Next came the sheep. There were several days of herding and cutting in random directions until it was clear to Tater he was dominant. The sheep never doubted it. Finally came the tractor. Several days of racing to cut the tractor off and block its passage until finally, it was clear. Tater was king. God save the king. And for good measure, God save us all.

The chaos of a farm can make one forgetful . . . as can turning fifty, if I'm truth telling. And, as noted, one of my bad habits is to occasionally leave the water running in the horse trough. One hundred feet of pipe can fill with air in about the time it takes to bring in sheep and horses; feed and secure them; collect chicken eggs; lock the hen house; start dinner cooking; and then . . . oh damn, I left the water on. Down I trudge to the pump house in the cold and dark. Crouching in the mud, face pressed against the water tank, staring at weird bugs caught in the beam of my flashlight, while I bleed air from the pipes. You'd think that the consequence would be enough to ensure that I'd remember, but apparently not.

So I was not surprised when I found Tater and Mora one morning, in the center of the barn, happily munching a bale of hay. The stalls were wide open. Obviously the old woman who ran the farm forgot to bolt the gates. When it happened a few days later, I became suspicious. Was I going crazy or was someone driving me crazy? Greg was automatically eliminated, since he had no appetite for running the farm alone.

My suspicions turned toward the new arrival. Was it coincidence that the gates fell open only after his appearance on the farm? And then there was the dubious delivery by the previous owners. Did they know more than they had disclosed? And here we were, one week beyond the return date. I shot Tater an accusatory look. He returned my gaze with an expression as innocent as it was dumb. Yes, dumb as a bag of taters.

I inspected the latch, which required dexterous fingers to slide the bolt over the buckle and out of its housing to gain release. No way could a horse open this latch with the bolt down. I looked at Tater. Especially not this horse.

That only left one other possibility. I was losing my mind. And the consequences were serious. The horses could get into the grain bins and colic.

I decided to double latch. I put a spring clip through the buckle making it impossible to slide the bolt out. That eliminated the horse theory for gate failure. The extra step of clipping the latch would force me to pay more attention. That eliminated the crazy old lady theory for gate failure. And it worked. I began to sleep through the night.

One morning I approached an eerily quiet barnyard. There were no horses in the paddock. Worse, there were no sheep in the loafing shed. Where there should be animals, impatient and demanding, there was emptiness and silence—until I opened the barn door. Over the beat of house music, horses and sheep scrambled like flappers fleeing a police raid. Bales of hay were strewn across the floor. Shovels and rakes and halter and ropes lay scattered about. Everywhere I looked there were large deposits of steaming party poop, announcing a good time was had by all. Fortunately the grain bins were still secured, but the teeth etchings on the lids suggested I had arrived just in time.

After getting all the animals separated and secured, the forensic investigation began in earnest. The spring clips had been flattened into pliable sheet metal by an animal with a jaw like a vice. The bolt had been lifted and slid out by an animal possessing an opposable tongue. Further inspection of Tater's stall revealed all manner of wood and metal objects, including live electric wire, had been chewed in frustration—proving, where there's a will there's a way. And apparently, Tater had been up all night "willing" the gates open. The "dumb as a tater" thing was just an act. Houdini would be a better name.

I upgraded to clips with teeth-breaking brass casing. Pricey, yes, but worth a good night's sleep. The next morning, as I sipped my coffee, I watched fluffy white balls munching their way across the front lawn. I had a moment of deja-vu, a feeling of our first days when a bright-eyed couple had just bought a farm, before the laws of causality caught up to us. Tater. He had to be behind it.

Sure enough, having been denied access to the barn, Tater, the freedom fighter, released all the livestock into the fields. I had not expected Tater to expend the great effort it takes to open a latch to a field he was released to every morning. Wrong. Not only would he expend the energy to open one latch, he opened all the latches. For Tater, the only good side of a fence is the outside.

Now all the latches on our farm would have to be double latched with heavy brass clips. While that would provide security from Tater's midnight raves, it came at a price. Not only were the brass clips expensive, they were cumbersome—especially for one-handed Farmer Jones. Initially this resulted in quite a bit of grumbling, until I reminded my spouse that he was the architect of our current discontent. It was he who had brought Tater the Terrible into our lives.

Truth be told, we both secretly admired the cleverness of this horse. Tater was a presence on the farm that could not be ignored.

Greg stopped grumbling and instead took the Tater challenge: design a latch that people could release but Tater could not. Ideally, a latch that could be opened one-handed since most farmers had something in the other hand when opening gates. After some head scratching, he hit upon a chain that would drop into a V slot to latch. Tater's tongue could push a rigid bolt up and out, but he couldn't push a limp chain down and up again. The latch was simple to use and easy to make. And it had a second slot for double latching. It was both simple and elegant in its application. Tater the Terrible had been tamed. He soon gave up his drive to drive me crazy.

And if it worked on Tater, it would work on other Houdini horses.

Viola! There it was, the invention that could save our farm. Simple design, easy to use, and if we could produce it for a good price, other farmers would want it. This is where Greg handed it off to me to do the detail work of getting a patent, a fabricator, and a distributor. In other words, the hard work. That's okay, hard work is what farming is all about and if it can both secure our livestock and secure our financial future, I'll do it. You bet I will. After all, this is farming. Born in optimism, deflated by hardship, resurrected through innovation. This episode started with a gift that turned into a problem that morphed into a challenge that produced a completely unexpected outcome. I guess the lesson is to stay with it long enough to see it through to the other side.

But I wondered, can you stay suspended in frustration too long while awaiting a solution that may not come? Is there a point where tenacity just becomes stubbornness? And how would you know the difference?

For the moment, I had a latch to sell that was Tater tested. A good latch, and a good chance to turn this farm around. And I had a clever horse that was secure in his stall. And I had a husband who thought to ease my season of loss with a little gift of redemption.

AWAY, YOU WAYFARING BROTHERS

The Wayfaring Brothers seemed to possess all the necessary ingredients to be successful entrepreneurs: tools, a reliable truck, and an impressive skill set that filled a need in the community. They were both self-reliant, requiring no oversight to get the job done, and dependable. Most importantly, with Jack in the lead, relations with customers could be managed and, when necessary, massaged.

The only snag in their path to financial success was capital. They lacked money. Well, that and they lacked the motivation to acquire money beyond their daily needs. Both the boys were practiced at the art of essential living and nearly immune to the lure of money. Especially when it came with conditions, as it so often did.

This was readily apparent in their living arrangements. Luther lived in a bunkhouse behind his mother's place. Jack rented a corner in his daughter's barn. Amid the hay bales, he had cleared a space for his cot, a table, and a trunk for his duffle. His only companion was a feral tomcat who recognized in Jack a kindred soul.

Like his companion, Jack accepted handouts from a number of women in town who secretly hoped to domesticate him. And like his companion, domestication was not in his nature. In the morning he was always gone. And though many were left disappointed, none was surprised, and most left a bowl of milk and a key on the back porch. Just in case either of those ramblers changed their minds.

It was easy to understand why women liked Jack. After all, Jack liked women. He loved the feminine wherever it presented itself, which was probably why he chose willow as his favorite medium in his craft. He planted willows all over his daughter's farm and harvested them for weaving baskets, bending into widgets, and building deck chairs. In the willow he found the supple strength and curved contours that reflected the beauty of nature. Of course, Jack saw it in simpler terms. He loved willows and he loved women. He didn't know why women loved him, but he was grateful whenever they did.

Then too, Jack would add, fortune just tended to find him.

On a bright summer morning, a semi picked up a little too much speed to safely negotiate a corner, causing it to swing wide of its lane. The truck in the oncoming lane began blowing its horn, startling the semi's driver and resulting in his over-correction. The semi fishtailed and its trailer slid off the road and into Jack's yard. Half of a manufactured house lay upended directly in front of Jack, who was taking his morning cigarette with coffee. For the driver, the load was a total loss. Worse, he would have to get a crane to collect it for demolition. Jack, with an air of magnanimity, offered to clean the whole mess at no cost. He smiled, the driver smiled, they shook on it, and the deed was done.

Jack used the farm tractor to hoist the house onto blocks. He hammered out the dents and nailed plywood over the gaping openings in the half-house. And with that, Jack had a new house. Like manna from heaven.

Of course, a house is not a home until you move your furnishings into it. An hour later, Jack had a new home—well, almost a new home. Every transition contains a seed of unexpected change. The cat declined the new digs. Hard to say why—fewer mice or maybe just a little too domesticated for his taste, but his refusal was absolute.

At the same time Jack was going through this break-up, a black Lab named Rosco was having a relationship disconnect of his own. Nate watched from his perch behind the counter of the Merc as a BMW pulled into the gas station across the street, let Rosco out, and pulled away. It was far too common an event. Surging with disgust for his fellow man, Nate walked across the street and caught Rosco by his collar. He tied Rosco to the bench in front of the Merc and placed a sign that read FREE TO GOOD HOME. Jack walked by and suddenly realized he had the necessary qualifications. And that's how a black Lab came to occupy the cat's place in Jack's new home.

It was a good fit. Free from the barn, Jack created an outdoor patio space with a fire pit and a grill. On summer evenings, he and Rosco would sit on willow chairs, next to the fire, waiting for their food to cook. Jack would play the fiddle while they sipped beers. Rosco's collar was replaced with a bright bandana that matched Jack's Mexican belt. Given that they both possessed the same disarming smiles, the effect of seeing them together could be unnerving. This was especially true when they were bunched together riding in the truck. Some said that when Jack wore his hat and dark glasses, the only way to tell them apart was to wait for one of them to speak.

So on a mid-summer morning, the Wayfaring Brothers, plus Rosco, took hammers in hand and began constructing a house for Annie. It had been a long process of pulling permits, milling lumber, digging septic, scraping the pad, and laying the concrete foundation. Finally the boys were here and the walls were about to go up. Luther took measurements, walked the site, took more measurements, consulted the blueprints, and finally directed Jack to start cutting. Jack laid the boards out and turned on the saw. Luther turned it off. He inspected the blueprints, took more measurements, walked the site, a final measurement, then nodded to Jack. As Jack began to feed the board, Luther ordered a stop and repeated the entire process a third time.

Jack erupted in exasperation. Luther explained he saw a way to save saw cuts but wanted to be sure. Jack replied whatever was saved in saw cuts was wasted in re-thinking. "Maybe," Luther allowed, but he wanted to be sure. Better to just do it and fix the mistakes, Jack argued. Luther relented, "Go ahead, cut." Jack fired up the saw. "Wait. Stop!" Luther had reconsidered. Jack cut anyway. Once committed, Luther proceeded down the path selected by the saw toward a productive day.

As they were storing their tools, Jack thought to put a conclusion to their disagreement. "Good day. We got a lot done without too many mistakes." He tried his smile on Luther.

"I just like to be sure," was the response.

Rosco stayed out of it.

The next day went well, but the week saw a reemergence of the conflict. It was an argument that would not go away because of the hardwired nature of each man. Luther lived behind his mother's place, not a hundred feet from where he was born, because he liked predictability. Jack lived in a barn because he liked the freedom and spontaneity it offered. If you don't like a room, then move the bales. They both enjoyed lives of minimalism but for very different reasons. When things were working, their differences complemented each other. Jack created the inspiration and Luther provided the regulation. But when things fell out of alignment, the synchronicity seized up like an engine with pistons firing against themselves.

As the summer progressed, Luther took longer soaks in the tub before responding to Jack's pounding at the door. The more pressure Luther felt, the slower he moved, and the later they showed for work. By the end of the week, it was Jack leaving early to go fishing and reset his frustration thermostat.

Despite their conflicts, the Wayfaring Brothers maintained fundamental respect and affection for each other, causing some to note the similarity of their fussing to the bickering of an old married couple. This caused others to point out the irony that these confirmed bachelors had snared themselves in the same net they had so scrupulously avoided with women.

It caused me to note my house wasn't getting built. And then I remembered, this is the country. I could hire a construction company from the city, but with the commute they wouldn't get here any earlier—and I would be charged for the travel time. At least it was Jack, not me, pounding on Luther's door. Maybe a city construction company could knock the job out, but I think I would miss all the idiosyncrasies that came with the Wayfaring Brothers. Like Jack whistling to my turkeys and the turkeys gobbling back. And Luther crabbing about being stuck with a turkey talker for a partner. And Rosco jumping into the truck's cab to blow the horn at quitting time. Doubt I would get that with a city construction crew.

It was a Friday when Jack's truck arrived without Luther. Being the last day of the week meant it was the longest soak day for Luther. My assumption was that Jack's patience had given out and he had come to work sans Luther. Then I noticed there was no Rosco either and I felt a twinge of alarm. Then I noticed Jack and I felt more than a twinge. I had never seen him without his huckleberry smile.

"Someone shot Rosco last night."

"What? Someone? Why?"

"He was barking and they shot him and drove away."

"Drove away in a Dodge pick-up I bet. That's Rodger Gates not wanting anyone, especially his wife, to know he's been seeing Melissa Johnson."

"Maybe. Don't know for sure and don't really care about the who, what, why, whatever. There's no cause to kill a dog on his own property."

"No cause and not legal. Did you call the sheriff?"

Jack just shook his head. "Karma can take it from here. I'm going fishing. Got some dark thoughts rolling around in my head. Need to get out on a lake. Just me and a fishing pole and a lot of still water. I'll be back Monday." He banged the side of his door with a two-beat farewell and drove off.

Monday he was back at work with Luther and with his smile. But there was no turkey rap or storytelling or anything but the staccato pounding of hammers and the forlorn screech of the saw. And there was no mistaking the empty spot on the grass where Rosco use to lie, waiting for the men to finish their distractions so they could attend to the priority of playtime with him.

Into empty space, nature always sends a tentacle. A root burrowing in, pushing out earth, making way for something new. So it was no surprise when Jack's truck veered off the highway and into the driveway of a farm with a sign FREE TO GOOD HOME. This time it was a burro, or a roll of shag carpet on legs, depending on the point of view. The owner was frustrated by the animal's stubborn temperament. Purchased as a pack animal for hunting trips, the beast, named Hershey, had proved intractable. Now it just cost money in upkeep and created problems with the other animals. Sooner it was gone, the better.

Jack thought the burro would make a great advertisement for selling baskets. He could mount his willow baskets on either side of the burro and take pictures. In his imagination, people would flock to buy them. Personally, I thought Jack was just a sucker for orphans.

The owner told Jack he would need a horse trailer and possibly a cattle prod to load this son of Satan. Jack backed the truck up to a dirt berm, threw a halter over the burro, walked him onto the bed of the truck, and tied him down. He thanked the dumbfounded owner and drove off. Halfway home he passed one of his favorite taverns and pulled over. In a celebratory mood, Jack invited the burro to join him. After relating the story to the patrons of the bar, it was agreed the burro probably adopted an incorrigible attitude in response to being named Hershey. There was a contest to select a more appropriate name, with Paco winning. They offered Paco a beer to test whether he liked the new name. He did. And so Paco now joined Jack around the evening campfire, happily munching grass while Jack played his fiddle.

But Paco was not Rosco. He couldn't join the boys at work. Jack worried that Paco was lonely, so he traded a day's pay for a lamb to serve as a companion. He also wanted the lamb as a gift for his granddaughter. It's a country fashion to give children lambs to help develop their sense of responsibility. So, I offered Rusty, one of my bummers.

Generally bummers are a good choice for children because they are people friendly. The down-side is that bummers are more prone to escape. Having lost their fear of humans, they are often watching how gates are being latched when the other sheep are running away. Rusty broke out the first week at Jack's place.

Jack spent most of the night walking the highway looking for Rusty. Besides the threat to the lamb's life, there was a significant risk to any driver that might hit the lamb. A hundred pounds crashing through a windshield could be fatal. Nothing turned up the first night. The next day, Jack shuffled into work reporting the lamb AWOL and possibly headed home. After several more days without any sightings, it was presumed Rusty had a cougar encounter.

On a slow Saturday afternoon, Rusty came prancing through an open door at the Merc. Several people lunged for him, which caused him to bolt down the aisles, knocking down displays, scattering cans, and breaking bottles. He circled back and leapt out the door, into the street. Cars swerved, brakes squealed, more chasing until he ran back into the Merc and completed another loop of wreckage. The whole town pretty much turned into a sheep rodeo, until Cody, the local farrier, happened by. He possessed a lariat and the skill to use it. In short order, Rusty was tied to the bench in front of the Merc. A call was placed to Jack.

Jack described it as the walk of shame. Half the town had lined up to see what kind of wrath Iron Nate was going to unleash on Hippie Jack. What most did not know was that both men were Marine combat veterans and the Semper Fi brotherhood more than covered this slight indiscretion. There was no scolding. Nate handed Jack the rope and then helped him carry Rusty back to the truck when the lamb refused to walk. Jack drove away. Nate did ask several people if they intended to buy anything, and if not, maybe they could occupy space in their own damn homes.

Jack took quite a drubbing from the town folk for a few weeks. A lot of whistling "Mary Had a Little Lamb" and questions like, "Do you know where your lamb is tonight?" For the locals, conditioned by lives of hard work and careful management of scant resources, the idea of getting a lamb as a companion for a useless burro seemed preposterous. The lamb escapade just seemed the perfect vehicle for shining the light of ridicule on this frivolous behavior.

I knew differently. I knew the lamb escaped after Jack's granddaughter failed to secure the gate. I knew Jack never disclosed this fact to anyone, including me—I learned it from his daughter. I knew the burro was a replacement for Rosco. That Jack, rather than focusing on retribution, had chosen redirection. I also knew that Jack lived a life of hard work and scant resources too, but rather than be conditioned by it, he chose to challenge it. He chose his companions precisely because they were impractical. The burro was both an act of defiance against the mundane and an embracing of mirth and folly. One look at Paco, with his big ears and oversized head mounted on a wine barrel body with stubby broomsticks for legs, would tell you that.

So Jack bore the ribbing with his trademarked smile. When confronted with the burro as yet another misstep in a life of stumbles, Jack responded with gusto, "Some mistakes are just too much fun to only do once."

I think that is how I will always remember Jacky. That and his smiling, ubiquitous salutation, "Bright moments to you." And his unflappable "We can do that" answer to problems, which always shattered whatever spasms of doubt and defeat I was experiencing at that moment. Yes, that is how I remember our Jack.

He was found dead on a Monday morning, sitting in his willow chair. A cup of cold coffee in one hand and a burned out cigarette in the other. It was the last week of our house construction, just the finishing touches where Jack would have left his artistic signature. He was feeling satisfied with the long project and looking forward to taking time off. So it surprised me a little that he was late. I assumed there were problems with motivating Luther. My surprise transformed to alarm when Luther answered the phone saying he was still waiting for Jack. A call to Jack's daughter sent her over to his house and the terrible discovery. We will never know the cause, but heart attack seems likely.

On a Wednesday threatening rain, I helped dig my friend's grave in the heavy clay soils of the Coast Range. Jack died with twenty-seven dollars to his name, but he had a wealth of friends who came together to get him buried. Some washed the body; several built the coffin while others dug the grave. Wherever people gathered, there were stories. The stories he told, the stories he lived, the stories he gave to us.

Jack was an artist of folk. Brick might be our mayor, our community organizer, but Jack was our artist, our weaver of tales. If you needed a tool, Brick knew who had it to lend. Jack connected us through our stories. He shone a light on our stories and let them find their connections within each of us, to each of us. It wasn't a connection to a community so much as a connection to a deeper level of ourselves—to our commonality as "folk." It was a "folkway," a river of lore, and he was our ferry man.

None of us were prepared for the loss. We had all expected to share many more years with this carefree soul who saw life for its opportunities and bright moments. Such a shame to lose his presence, to lose his light, to lose our way a little bit, out here in the country.

THE COMING OF AN END

Luther took the loss especially hard. Jack had turned Luther from his reclusive ways and pointed him back into the community.

Under Jack's sway, Luther began to show up at social events and reengage old acquaintances. Where before he was known for his eccentricities and social aloofness, now he was recognized for his quiet intellect and sardonic wit. Jack's passing reversed all of that. Luther was rarely seen off the family farm, and even there he stayed in the shadows.

After several weeks of Luther's absence on the job site, I became anxious about completion. With only a week left, it just needed this final push. I called and left messages and pestered him back onto the job. He worked in piecemeal fashion, always grumbling. The week of work stretched into a month. One day I caught him crying. The turkeys, seeing him at the house, had called to him, triggering a memory of Jack and his whistled response.

Until that moment, I hadn't quite realized the burden I had placed on Luther. I had slipped into thinking of Luther in that one-dimensional way of stereotypes. The grouchy hermit without a heart. Sitting with Luther on the steps, the house full of painful memories, even I could see Jack coming around the corner at any moment and then I was tearing up too. I sent Luther home. Greg, or someone, could finish the little bit that needed doing. It was the last work Luther ever did off his farm.

A few weeks before Thanksgiving, on a frosty November night, coons got into the coop and slaughtered all the turkeys. Blood lust, I guess, because once they started killing they didn't stop. I felt something inside me break. All this loss, and now the turkeys. I was depending on the turkeys to get us through this year. It wasn't killing to eat, it was reckless, empty slaughter and it just seemed cruel.

Thanksgiving brought the rains and with it the migrating salmon. There were days when my spirit lifted, watching these magnificent fish, these harbingers of bounty. But the rain didn't stop. Torrents of water ran off the mountain slopes as nature seemed to rage at everything in her creation. The floods lifted bridges and washed out roads. When, at last, the rain relented, we found salmon eggs spread across hay fields. I could only shake my head at the squandered loss that lay at my feet. I scooped up what eggs I could and placed them back in the stream, but it felt like an empty act against the predominant theme of destruction.

I was beginning to feel like I was being served an eviction notice from mother nature.

As Christmas approached, our girls returned home—or rather to the farm. This year our small Christmas felt small. It was less rustic and more just hard and cold and pinched.

I challenged myself—there was reason to be thankful. Despite all the loss, Annie's house was built. That overt promise of legacy and renewal. It was a beautiful little cottage with majestic views over the fields and the panorama of the mountains. Yes, it had been a bad year, but the house alone, with its promise for the future, could offset all of that. If next year was a good year . . . and with the cottage . . . and Annie coming home . . . it all could be resurrected. We could turn the corner and make this farm work. I couldn't wait to show it to the girls.

After viewing it, Annie pulled me aside, "Mom, the house is beautiful, but I didn't ask for it. I'm applying to veterinary school. If I get in, I'll be gone at least four years. And truthfully, even if I don't get in, I don't see myself living here. It's just too isolated for a single person with a social life."

I was struck dumb, averting my eyes to hide my disappointment. To distract myself, I looked at my other daughter. She immediately threw up her hands in defense. "Don't look at me. You know I can't live more than a mile from my manicurist."

Well, at least I raised daughters who can speak their truth.

What was it I said about farms? Oh yes, they're the destroyer of delusions. Time to face facts. It was over and Greg and I both knew it. Even if we succeeded in making the farm self-sufficient, to what end if there was no legacy? And the accounting this year was going to be the worst on record. We increased expenses by building a new house but lost income with the turkey genocide. And the gate latch, where's the money going to come from for fabrication and distribution? There was no reason to believe that the farm could be self-supporting. Not this year. Not any year.

PART FIVE

FINDING THE QUIT IN ME

It was time to quit the farm, but quit it how? It's as hard to quit a farm as it is to start one. There are two basic methods to accomplish a quit. The first requires admitting defeat and then developing an action plan to divest. A call to a good realtor will get you started. I opted for the second method, which involved pulling the covers over my head and going back to bed. Bubba, in a rare moment of compassion, joined me, curling up at my feet. Out of the darkness came the consoling, rhythmic hum of a purring cat.

I tucked into a good rumination. After an hour of counting my losses, I attempted a last stab at problem solving. I systematically reviewed the four M's of farming (money, market, manure, and machinery) hoping one of those categories could offer some, previously overlooked, reason for hope.

The first M—money—is the most crucial part of farming. It takes money to buy the farm, keep it running, and to provide a cushion when things fall apart. And things always fall apart at some point. For us, they had been falling apart at an alarming rate. We had enough money to afford the farm if it was a static entity, but it wasn't. We were continuing to bleed money. So the first M—problem unsolved and soon to be insolvent.

Ideally, the problem of money can be answered with the second M—market. The right market can increase profit and staunch the flow of capital waste (pay debt or replace antiquated equipment). And in farming, there are only two solutions to market: scale up (go big) or find a niche. Scaling up was out of the question. First, we lacked the capital to expand. Second, we picked the one ag market with a steady decline of sales. Americans have been eating less lamb and wearing less wool every year for decades. That's not a trend showing much promise of reversing.

The only answer was to mine the niche market, and really, our lambs were a niche. But I had not been successful at creating a demand for Katahdin lamb. Heritage turkeys? Maybe, but the latest coon attack left me dispirited. Up in these woods, predation was a constant, and turkeys just seem to invite their own demise. It would be a constant fight with mother nature, and to date, I hadn't fared very well with those battles. I wanted an accord with that good mother, not a battle. So the second M—problem unsolved.

The third M, manure, or more accurately soil quality, was not in our favor either. Good farmland is found in river bottoms, not mountaintops. We chose pretty over productive. The soil quality greatly limited what we could grow, and the rain kept the good soil and nutrients moving downhill. We were working at such a disadvantage that even with composting and good management, the best we could hope for was to stay even rather than improve the soil quality.

And our soil just couldn't compete with the productivity of soils in the valley below us. At one time my little community boasted a sizeable farming population to feed the sizeable logging population. Now logging is largely mechanized so the population is small, and improved transportation puts me in direct competition with my flatland neighbors. It was a losing proposition. This simple fact of geography and its influence on farming is one of the great drivers of world history, yet it never even crossed my mind living in the city and shopping at my supermarket. The third M —problem unsolved and probably unsolvable.

The fourth M, machinery, is at the very center of farming. Farming began with tools, was the result of tools. Ten thousand years ago somebody came up with the idea of a scythe. Before that, people walked up to wild grains and just shook the stalks, gathering whatever seeds would release into baskets. Some nameless individual chipped tiny flakes of razor sharp obsidian and imbedded them into a stick; thus creating a tool (scythe) that allowed people to cut the stalks. The stalks were then taken back to the village where all the seeds could be shaken out. Probably the intent was just to increase the collection of seeds, but the result was that seeds that clung to the stalk were more likely to be harvested and ultimately to be replanted. Over time, the scythe produced a variety of plants with seeds that don't randomly drop but rather "wait for the harvester," or, as we like to call them, domesticated cereals. Agriculture was invented, and human history was transformed right along with the cereals. Sheep would follow shortly.

Tools define farming, and the right tools define success in farming. Cash-strapped farmers are often working with worn out or obsolete machinery, hoping to patch and keep them functioning. Of course, the obsolete equipment is less efficient, lowering the farmer's profit margin, and is more prone to breaking down, leaving the farmer in the bind of paying for a repair, with less income to cover the cost. Or the farmer can go into debt to get better equipment and hope that the harvest will cover the cost of the investment. We purchased a tractor, which helped us to decrease the cost of our feed but, in the short run, it increased our debt above what it saved us. Over time, it should pay for itself, but for now it was a drag on our budget. We could make other investments in equipment that, long term, would lower our cost and improve our efficiency, but we lacked the capital to do that. So the fourth M—problem unsolved.

So there it was, all four Ms a total bust. And let's not forget—no legacy. No one to take over all that we'd built—or, more accurately, to live in the hole we'd dug. The only questions remaining were how to sell the farm and where to go. Once again we would pull up our lives and move to . . . what? We'd have to go to a city where there were jobs, but what city? What jobs? I burrowed deeper into the covers.

Downstairs I could hear a door open and then the slow heavy trudge of work boots on the stairs. Greg was calling me. At the bedroom door he paused and then stuck his head into a totally darkened room. Observing his wife curled into a fetal position under a cocoon of blankets, he asked tentatively, "Are we out of milk?"

This from a psychologist!

"Probably."

Apparently my tone carried the right amount of threat. He muttered "Okay," and closed the door, leaving me to my funk. I think I was hoping for one of those George Clooney moments where your spouse, using tough words in a tender voice, coaxes you back to a world full of promise and hope. What I got was my husband lost in his world and not quite knowing how to find me in mine.

And why should he be Clooney for me? He probably just came in from a whole host of farm problems for which he had no answer. Maybe he was hoping to come home and sit in the kitchen and have a wife serve him a glass of milk with a slice of pie, while nurturing him back to a world full of promise and hope. What he got was a woman barricaded in her blankets.

And that was all either of us was going to get that day. That was all we had to give. The farm had taken the rest.

As he closed the door, it went dark again. Bubba started his droning purr. Not exactly George Clooney, but it helped remind me I wasn't alone. I began to reflect on how intuitive animals are about emotional needs, even cross-species. Here was Bubba, the brat, sensing how much I needed his purring at this moment. I began to relax with that warm thought and stretched just a bit under the covers. As I did, Bubba clamped fang and claws on my big toe.

That got me out of bed. Bubba sprinted out the door faster than I could throw a shoe. I sat on the end of the bed considering whether to go fetal again, but to what end? The farm would just call me back to this reality. It could match every moment of moody self-contemplation with an imminent call from the crisis du jour. So, absent a plan or a vision, I did what farmers do. I did my chores.

I even fed one despicable black cat. But only after I fed the dog first. Revenge served cold, or at least tepid. Oh, and I also put in a call to a realtor.

SPRING REVELATIONS—MIRACLES FROM THE MUD

During this time of darkness and fog, we were visited by a mystery. Two lambs were born without a mother. This was more spontaneous combustion than immaculate conception.

We had just started lambing. At birth, the ewe-lamb combos are moved into the barn for three days of bonding and tagging. Following this, they're moved to the lamb pasture with extra rich grass for the nursing ewes. Thus, the herd is divided between expectant and delivered ewes.

So what were two new lambs doing in the already delivered pasture on that crisp, spring morning? It is possible to have delayed births. A pair of lambs arrives and then, almost as an afterthought, a third appears a few hours later. Delays had been reported up to eight hours, but not days later. All of the ewes in the field had lambed more than a week ago, making a delayed birth impossible.

The answer had to be a sneaky, fence-squeezing ewe in search of greener grass. Obviously she had pushed into the field, birthed, and then pushed back out, leaving her lambs behind. Not good maternal behavior, so probably a new ewe. A few days in the barn with her babes should repair the attachment disorder. The challenge was to find the wayward mother and reunite her with her babes. And find her fast.

Time was of the essence because lambs need colostrum. There is about a forty-eight hour window before the lambs' stomachs would "turn off" and not be able to absorb the important immunological benefits in the colostrum. Without the colostrum, they were a high mortality risk. Time was also vital for the bonding. Without the stimulation from the lambs, the ewe's body would begin shutting down the production of oxytocin necessary for bonding and lactation. The lambs had been cleaned, which stimulates the production of oxytocin, so the ewe should be bagging up and have a real interest in lambs. It shouldn't be difficult to find her; biology would be pointing at her like a huge neon sign.

Except there was no flashing sign. All the old ewes were obviously still pregnant. Two new ewes were suspicious but new ewes often have underweight babies and hard-to-detect pregnancies. Neither of the new ewes was bagging or showed obvious signs of recent births. My skepticism was confirmed when they both went on to deliver the typical scenario of underweight, single lambs at a later date. So, none of the ewes in the expectant pasture was responsible for the lambs either.

The only possible explanation was spontaneous combustion. Some mixing of gasses from decaying matter and maybe a static spark. Greg offered a time machine theory, which was ridiculous. Their tiny hooves could never reach the levers.

My immediate concern was getting colostrum into these babies. In the absence of a mother, I could milk a substitute ewe. I needed a ewe that had just given birth, so I went to the barn. Except, there were no ewes in the barn. The ram had apparently taken a few days off last fall to regain his strength. Plan B was to buy powdered colostrum from the feed store, except they were out. Plan C was to call a neighbor sheep farmer. Experienced sheepers know to milk ewes for extra colostrum and freeze it for just such an emergency. Except, none of my sheeper friends had reserves. I hoped for a new lamb delivery in the next twenty-four hours, but none came.

This was spiraling into another hard-luck farm story. It starts as a miracle but ends in a loss. I brought the lambs into the infirmary kitchen. Sitting on the floor next to the wood stove, I pushed nipples with warm milk into their mouths. Once I got them started, I could switch both bottles to one hand leaving the other hand free to stroke their butts, just as their mothers would do. They began sucking but couldn't put that together with swallowing. The milk dribbled out the corner of their mouths. Cisco began licking their faces with delight and, with every lick, developing his own attachment to these lambs.

I was crouched on the floor, with two lambs straddling me and a dog in the middle, when Greg came in. I couldn't greet him because I had one hand on the bottles, one hand on lamb butt, and a dog in my face. He took off his coat and gave the scene a long gaze of assessment. He already knew the basics. They were bummers and, without colostrum, they were doomed. So what was I doing feeding them?

What indeed! I didn't have a good answer to that. In fact, I didn't even like the question. And I was preparing to not like the person who would ask that question. Apparently he missed my cues, because he just waded right in.

"You know it really is a shame."

"Yeah, what's a shame?" So here we go with the long windup.

"You truly have a farmer's heart."

"Meaning I'm either stubborn or stupid, I suppose."

"Well . . . we agreed we're calling it quits, right? Putting the farm on the market? It's over. We're done. Finished. Probably sell the herd at auction for pennies on the dollar. Two more lambs? Makes no difference at all. None. And that's if they live, which they won't! Yet here you are." He was shaking his head.

"Un-huh." This guy was really not paying attention to the emotional currents in the room. He just kept wading in deeper.

"I was the one that wanted this lifestyle, I know." He raised a hand, as if to take responsibility. "But you're the one fighting for it. You've got the heart for farming. Not many would have stuck this out, and look at you." His eyes locked onto mine. They were intent. "Even now, knowing we can't make it—no way, no how—you're still farming. You got some grit in you!"

He paused to let the full weight of his words register, then continued, "Shame to discover that now, as we're leaving . . . I guess, sometimes . . . I'm just slow. But I'm grateful to know it. Makes me love you all the more."

Wait . . . I think I'm having my Clooney moment! I just sat there. Lamb sucking. Dog licking. Me stunned.

He started to walk away, but I called out, "Hey!" and nodded to him. He walked back. Neither the lambs nor the dog were forfeiting any space. Over the sound of sucking, I looked up and smiled. He bent down and just rested his forehead against mine for a moment before taking a bottle with a lamb attachment.

Fortunately, I can count on Greg to not linger in Clooney similarity too long.

"So I'm guessing dinner is delayed tonight, 'cause I'm third in line after the dead lambs and the dog?"

"What makes you think you're third? Get me the tube. These lambs are sucking but they're not swallowing. And they're not dead. Not yet."

They weren't dead. And they didn't die. I tube fed them until they could swallow. I tied off a ewe and milked her, so they did get some colostrum. Maybe it was enough or maybe they were just tough. Whatever the reason, they lived, and I needed that miracle right then. So perhaps it was more immaculate conception than spontaneous combustion. Some gift from above to give hope.

My daughter, now a vet, has an alternative explanation. She noted our old reliable ewe, Didi, had only one lamb that year, unusual since she had always twinned in previous years. And apparently she gave birth just as a new ewe birthed a single. Annie's theory was that the new ewe actually had twins and Didi stole the first while the mother was birthing the second. In this conjured reconstruction, I failed to notice Didi was still pregnant when I brought her into the barn. She delivered her two lambs a week later in the lamb field. She licked them clean, possibly even allowed them to nurse, then returned to the single to which she had already bonded.

This is what comes from an education in science. A tidy little theory that explains things using only the known facts. It lacks creativity and seems to cast her mother in an unflattering light. I'm sticking with immaculate conception. These lambs were a miracle when I needed one. Miracles are as much perception as they are the actual event. I ask you, is it not a miracle to turn my spouse into George Clooney for one shining moment?

And maybe that was the real miracle. There is a profound healing power in the simple act of recognition. For that one moment, Greg broke through all the role-encrusted armor we had donned, to see me. Failed farmers we might be, but at least I felt appreciated for all my hard work. And that made all the difference. I was able to let go of the farm without feeling personally flawed. And not feeling defensive about my personal flaws made it easier to appreciate how hard Greg had worked. And recognizing his hard work allowed us to let go of the farm as a couple, rather than flawed individuals, each secretly blaming the other for the failure. This would be one farm foreclosure that would not take a marriage down with it.

And that made me love him all the more.

OF DONKEYS AND DRIVE-BYS

Every day on my way to chores, I walked past the empty cabin and it weighed on me. It was Jack's cabin and I couldn't think of it without hearing the banging of his hammer, his turkey whistles, or his conflagrations with Luther. I couldn't see the cabin and not think of Jack. And that loss reminded me of my other loss—my dream of living with my daughter next to me. It was time to do something. A change was needed.

A house down the road was listed for sale. The agent would already be familiar with the area, so it seemed a good place to start. Tanya, the agent, told me the house had just sold and I would be getting new neighbors. Good, I noted, if she can sell my farm they can be new neighbors together. I had found the quit in me. Tanya came right over. Assessments of farms are laborious, so she wanted to get started right away. I liked Tanya.

She had just left when Brick drove up the driveway. In the bed of his pick-up was Jack's burro, looking shaggy, soggy, and thin. Brick had a big smile.

"Hey, I got a favor to ask."

"No."

Brick laughed. "I guess you have gone country. Every favor comes with a price and sometimes two, right? But not this one."

"By that you mean, especially this one, so the answer is especially no."

Brick laughed again. "No, seriously, can you hold onto Paco for a day or two, just until I can get a pen built?"

The truth is that any favor Brick asked could not be denied. Our debt to him was enormous and, while he might skip out to go fishing, he was always there when it counted. "Okay, so a day or two is how much in Brick time?"

"Forty-six hours. I'm giving you two hours as interest on the loan." Brick smiled.

"Let me ask Greg?"

Brick blew the truck's horn to call him. Greg came around a corner and took one look at Brick, then the burro, then me. "No!"

I decided to sit this one out. Brick cocked his head and laughed, "You know what they call a drive-by in the country? When you leave your car window down and someone throws a bag of zucchini in the back."

This might have been funny if it weren't true. Zucchini grow in abundance, and by the second week of August, you look for "friends" to receive the gift that keeps on gifting.

Greg was not laughing. Brick continued, unfazed. "It's only temporary 'til I get the pen built. I drive by Eva's place every day and see this poor animal out there suffering. This was Jack's burro. Eva's got her hands full working and raising a family. I promised her I'd give it a good home. Just hold him 'til I can get the pen built—couple days."

As burros go, this animal was worthless and pitiful too. It was a miniature, Sicilian burro, meaning it had a full burro body, including big ears and big belly, on stubby little legs. The Sicilian meant it had extra-long hair, which was matted and hanging over its eyes, with globs of dried mud attached everywhere. It resembled a walking haystack with a black nose.

Brick untethered the donkey and led it down to graze on the grass. Instead of grazing, it walked over to Greg and nuzzled against him like a big, gray dog. I had to look down to keep from laughing. This burro was a delightful con artist, not unlike Jack.

Greg began to stroke the hair out of the burro's eyes. "Two days?"

"Forty-six hours." I clarified, "Two hours for interest."

Brick laughed, "Probably less."

"'Cause if it goes one minute over that, I'm bringing this donkey down to your house and tying him to your front porch," Greg asserted.

"And you'd be in your rights to do that." Brick responded.

Then Greg turned to me and with the full measure of his moral authority, pronounced, "And don't you get attached, 'cause this animal is gone in two days."

Right, I'll be careful not to do that, Sergeant Rock, but then I wasn't the one scratching his ears at that moment.

Brick visited for a bit and then it was time to go. He drove casually down the driveway, careful not to hit the gas until he was out of sight. I lingered to watch him pull away. He didn't turn toward his house but instead turned toward the river. There was a fishing pole in the gun rack. No doubt, that's where he was headed when he came across Eva and her burro problem. Heavy is the head that wears the mayoral crown.

I doubted that we'd see Brick again for at least a week. We had been the target of a country drive-by. Instead of zucchini, it was a burro. I didn't mind because it was Jack's burro, imbued with some of Jack's personality. Brick understood this connection and was counting on it to make the adoption stick. Objectively, we were probably in the best position of anyone in the vicinity to take on a burro—that is, if we were staying. And it would likely take us many months to sell the farm and who knows what options would open for Paco in that period. Mostly, I was just curious how long it would take Greg to realize he'd been hit by a drive-by.

Of course, Paco still had to work out his place in the horse hierarchy. With his stubby legs, it was at the bottom. No use to challenge the obvious. If Paco needed to feel big, he could raise hell with the sheep, which he did occasionally. Paco's place was unique and he accepted it with grace.

As for me, Paco redirected me away from the cabin and all that loss. Paco came into Jack's life when he lost Rosco. Rather than focusing on the loss or retribution, Jack pivoted into a new relationship. The little burro was a reminder of Jack's light touch with life's hard problems.

We were going to have to pivot off this farm. I didn't know where to or even how to yet. But wherever we went, we would be taking our horses with us, and how much more trouble is a little furball donkey going to be? It just needs a certain lightness of touch.

The forty-six hours passed without incident—unless you count Greg rubbing Paco's ears an incident. A week later he noted Brick had still not built a pen. On reflection, Greg observed, Brick was a great guy but a bit "irresponsible" for the duties required of a donkey. After careful consideration, he thought it would be better . . . for the donkey . . . if we just kept him . . . until we had to leave.

I hadn't told Greg that Brick had already called asking how the "goat" was working out. I confronted him about doing a drive-by. He was a little chagrined but laughed and said it just made more sense given we had the setup. He added, "That little booger is cute." The truth is that Brick knew the deal was sealed when the burro nuzzled Greg. But I wanted to see how far I could take this.

"Brick agreed to forty-six hours and you've been more than fair. I'll put the halter on and you can walk Paco to Brick's front porch and tie him up, just like you said."

"I just don't think that's fair to the donkey."

"Deal is a deal. This is a farm. Every animal has to pull its weight. The donkey is not our problem."

"I don't know. I kind of like having the donkey here. Reminds me of Jack."

"You mean to tell me that you got attached?"

"Well . . . yeah."

"Then I think we should keep the donkey."

"What about Brick? Think he'll be upset?"

"I think he'll get over it. Why don't you take him a beer and see what he says."

That's what Greg did. Apparently, Brick has amazing recuperative abilities.

BEAR IN THE GARDEN, BEES IN THE TRASH

A yellow lab was making a nosey skitter across our front lawn. Trot, stop, sniff; trot, stop, sniff. This is bad behavior in the country and did not go unnoticed by our canine nation. They bolted out the dog door in full war cry. The lab realized his error too late to make an escape and did the next best thing. He rolled over in utter surrender. Both dogs were on him with ferocious snapping jaws but as soon as he went leg up, they pulled up and circled. There was some pawing and growling and a few snaps at his face, but the lab held his pose. Finally, the growls gave way to butt sniffs, and the lab popped up ready to play with his new friends.

With the canine conflict resolved, we moved on to the human conflict. This kind of bad behavior leads to dogs being shot. The dog had tags, I just needed to catch him and phone the owners. It wasn't hard since the dog approached me with wagging tail. After a proper introduction, I bent over to read his tag. Big mistake. I got a face washing like I hadn't had since that mud puddle incident at age four.

Through the licks, I was able to read that Louie belonged to one Mark Sunkett, residing in San Francisco. Licky Louie was a long way from home. Most likely Louie wandered from his owners while they were visiting somewhere in the area. Hard to know exactly what scenario brought Louie to my lawn and, ultimately, it was not my concern. This was a call to animal control. As I was making the call, another interloper—possibly a bear . . . with a bald head . . . and tie-dyed T-shirt, tanker shorts, and flip-flops . . . was meandering across my lawn calling for Louie.

I stuck my head out the door and called to Mark who responded with a quizzical look. "It's okay. Louie has already started the introductions. You're the new neighbor, right?"

The bear-man was quite genial, with large brown eyes and an impish smile—teddy bear was a more apt description. I invited him in for get-acquainted coffee and a muffin. He flipped open his cell phone to call his wife, Kathy, to join us. It was a real greenie move. I handed him the phone and reminded him we live in a slot valley. Then I demonstrated the proper way to call a spouse by giving a shrill whistle. Slot valleys may be bad for cell phone reception but they're great for echo. Greg and Kathy joined us before the coffee finished percolating.

Kathy and Mark had worked in the Bay Area as mental health counselors. Kathy, a willowy blonde with a voice so calm it could turn a Chicago trucker into a tree hugger, was a psychiatric nurse—a job in high demand with great geographical flexibility. Mark was hoping to transition into the construction trade. After being priced out of the Bay Area housing market, they came north to find a home with enough land for subsistence gardening. By lowering their housing cost and supplementing with gardening, they could survive on Kathy's salary while Mark transitioned.

So once again, urbanites looking to get back to the land. But they didn't buy a farm, only a large garden. And they were ten years younger than us with more energy to pull it off. So probably relatively sane. Still, they were in for a steep learning curve. We shared our story, minimizing key parts with the generic "it's a lot of hard work." They were still in the brightly deluded stage. It would have been cruel to throw that much cold water on their dream in our first meeting. And I left out the part about contacting their realtor to sell our farm. It would take Tanya months, and we had time to ease into that topic.

Instead we offered to introduce them to the community. They would need the formal tail-gate welcome from the mayor. We told them which beer he liked. And we told them they could get the beer and any other necessities such as ammunition, lamp oil, suspenders, ginseng root, bolt cutters, or pizza at Nate's Merc. And I invited them on my morning walk with Sarah, which they accepted with pleasure. We were in mushroom season, so our walks benefitted not only our muscles and minds, but supplemented our meals as well. Better than a gym membership.

This next part began with a mole. If you've never seen a mole, they're about the size of a sewer rat, with black fur, pointy snouts, sharp teeth, and bright pink paws that look like they're wearing gloves put on backwards. Their hair follicles are uniquely perpendicular to their skin so their fur won't bind when they reverse in their tunnels. They are a peculiar and rather sinister looking creature.

On this particular night, Bubba, our own emissary from Satan, dispatched a mole. He found it not to his taste and dropped it by our back door before moving on to more delectable mice. The mole carcass sat by the door the rest of the night and the better part of the next day before Greg disposed of it in the trash can. He did this in total disregard of the obvious omen posed by evil (Bubba) killing evil (the mole). I'm pretty sure that doubles the evil. You can fact check it in any witch's bible.

Having finished my chores, I was discarding a feed bag in the trash when several wasps emerged from the mole carcass, then several more, then a horde. Before I could get the lid down, they swarmed me, going for my only exposed region—my head. I swatted them away and dropped the lid but not before I was stung on the neck. I flashed on the intense pain from last summer's attack and the warning that I might be developing an allergy to wasp venom. I rushed inside for some Benadryl.

As I reached the door my entire body began to feel prickly hot. Not a good sign. Skip the Benadryl, where's the EpiPen? Yes, where is the EpiPen? I called Greg. By the time he arrived, I had found the pen. I checked the expiration date but it kept blurring out of focus. Or was it my legs wobbling? I handed Greg the pen and told him about the wasps with growing panic in my voice.

The psychologist took over. He began talking me down from my anxiety attack. I tried to relax but the symptoms weren't getting any better. "The pen. The pen." I remember getting that much out. My heart was racing, my throat constricting, and my vision was blurry, bright lights. The symptoms alarmed Greg and he called Kathy. He was better at talking than medicating. Kathy wasn't there but Mark came immediately. I was on the floor losing consciousness and the only thing between my death and the EpiPen was the deft reactions of these men. Greg called to Mark to read the direction while he administered the shot.

"EpiPen, manufactured for DEY, Napa, CA by Meridian Medical Technologies, Inc., a subsidiary of King Pharmaceuticals, Inc., Columbia, MD, USA."

"Skip to the directions."

"These are the directions."

"Yeah, just get to the important part."

"I'm trying."

"For Christ's sake hurry, she's turning blue."

"I'm calling Kathy."

And so they did. Kathy told Mark to hold the phone to Greg's ear. In her exquisitely calm voice she said, "Slam the pen into her thigh."

"What about her clothes? Should we cut a hole or something?"

Still calm, "Just slam the damn thing."

And so the boys did, saving my life.

The next thing I remember, I was lying on the floor, cushion under my head, and Greg dabbing a towel full of ice on my forehead. Mark was on the phone to 911. Kathy came through the door about ten feet ahead of the paramedics. Did I know who I was? Where I was? Yeah, it's still spinning but I could talk. They started an IV.

More people. The dogs kept pushing the door open and were chased out. Two hunky firemen lifted me onto a stretcher. Someone canceled a helicopter. More hunky paramedics loaded me into the ambulance. We sped down the gravel road past Brick pulled over in his truck. I remember the look of concern on his face. The rest was surreal. Facing backward as we whipped through the curves of the mountain pass. Kathy's voice so reassuring.

I spent the next few hours in the ER, throwing up. Gradually my heart stopped racing, my throat opened up, and the hives went away. The doctor noted it had been a close call and next time, not to delay the injection. Greg looked embarrassed. It may have taken him awhile but he got the job done when it counted. I gave his hand a little squeeze. There was Kathy and Mark hovering around my bed. I realized we had just forged a lifelong bond and I thanked them from the bottom of my heart.

Back home, we stocked up on EpiPens and reviewed the procedures for using them. We also laid in a good supply of wasp spray. I called Brick to reassure him. He was greatly relieved.

As for me, it was a terrifying event that traveled from the banality of farm chores to the cusp of death in minutes. Another delusion destroyed. That sense of safety, of time enough, that we all assume without a questioning thought. That sense is an illusion. We do not have enough time. And knowing that, what do you want to do with this life, right now?

Yes, what indeed!

RULES OF THE CIDER HOUSE, REVISITED

Having your life saved is an oddly ambivalent experience. While it's at the top of the chart for positive bonding experiences, it's difficult to assuage that persistent sense of debt. These people saved my life, after all. The Sunketts were gracious people and would gladly have forgotten the whole event. I, however, couldn't escape the feeling that I should end every sentence to them with pulsing gratitude. "Would you like a glass of wine and thank you for saving my life?" And perhaps it was their graciousness that kept that feeling in play. Greg, by contrast, was quite happy to have me eternally in his debt. I had to remind him that his contribution was to try to talk me out of anaphylactic shock.

Being gracious people, the Sunketts had a gracious solution to our friendship impediment. They proposed a juicing party.

Just like Bay Area hipsters to rename happy hour, I thought. But they had another kind of juice in mind. We had a small apple orchard behind our house that fed a variety of livestock and wildstock and occasionally a few people. They suggested we get together to juice some apples. They would supply the labor, namely Mark's bear-aptitude for shaking apples from trees. We would supply the resources: the apples and a press Greg had picked up at a yard sale.

So on a chilly November morning, we assembled on our concrete car port, with tubs of apples and a medieval-looking press, to juice. A bold move since our collective experience at juicing was to pay a store clerk for the finished product. Being children of the Industrial Age, we broke the task into a series of production stations. First, the apples were dumped into cleansing solution, then they were taken to a chopping station where they were quartered, then dropped into the automated pulverizer, then into the press. The press had a manual screw top the boys tightened down, squirting juice from the ribbed sides of the containment cylinder into a collection pan which was then funneled into a pure stream of liquid gold.

Easy enough, right? Except it was a chilly November morning. The cold eliminated the wasps that would have had a keen interest in our sweet nectar product. The trade-off was standing on cold concrete while retrieving apples from icy water baths, producing a numbing chill that makes the handling of sharp knives challenging. When the apples aren't quartered properly, they jam the pulverizer. Using bare fingers to free a pulverizing machine is never a good idea. Multiple thrusts with a wooden spoon work better, but that spits out regurgitated pulp that must be dodged to avoid having your face painted. And there's the compacted mash at the bottom of each squeeze. This has to be pounded out with a sledge while someone holds the press. It's inevitable that, at some point, the sledge will strike frozen red knuckles.

On this occasion, it was Greg's knuckles and the sledge blow was mine. Unlike the Sunketts, Greg is not a gracious person. He flung a slushy clot of apple mash at my face. Greg is also not quick. Certainly, not quick enough. I ducked and it hit Kathy in the face. Let the court record reflect that I was innocent of intent. She screamed a vulgar misrepresentation of my birth mother and threw a glob that hit me in the face. I responded with a particularly large, pulpy gob that hit her square in her flaxen hair. Yeah, that's right, sticky apple curds matting her flaxen hair. Apple juice probably dribbling into her ear. And for good measure, I plugged Greg too. Greg drew a bead on me but Mark nailed him before he could execute. Kathy came to Greg's defense and roiled Mark with a head splat. With the sides chosen, the battle was enjoined. There were no winners and no quarter given until we were mashed out.

I will say this. It is hard to look at your angel of mercy with apple mash dripping from her nose and think, *thank you for saving my life*. Personally, I think angels of mercy should comport themselves with a little more dignity.

The losers (we took a vote and it was the guys) had to clean up. We went inside to wash off the pulp and prepare a meal. We netted thirty gallons of juice, enough to see both couples through the year. But as with most things on the farm, the intended goal was only a process that lead to richer unintended consequences. Consequences that opened new possibilities with each other and with ourselves. We had made friends through a food fight. Hard to imagine that happening in my previous life.

FROM RUBBLE, MUSHROOMS RISE

Beneath the moist, green, life-giving canopy of the Pacific Northwest lies an abundance of rot and decay. And from this abundance of rot springs a delectable array of mushrooms—a reminder of the paradoxical duality of the life cycle. For, just as every living thing is destined for death, every dead thing is both platform and sustenance for new life in a continuous cycle. So on a beautiful autumn day, a common avocation in the Northwest is to tromp the sun-filtered woods in search of chanterelles or morels.

For farmers, the thought is, *why scavenge when I can grow?*

For the Sunketts this thought, along with consideration for the soft impact of mushrooms should they be hurled at your face, lead them to suggest we have a mushrooming party. The idea was to assemble suitable logs, drill holes in them, and then tap in wooden plugs soaked with mushroom mycelium. Six months later the log would sprout beautiful mushrooms from every surface, like kernels of corn popping from a cob.

Of course, not any log is suitable. They must be deciduous, and the heavier the grain the better. And they can't be freshly cut trees because the trees have their own immunological response against the mycelium. And they can't be lying on the ground because then they're likely to be infected with undesirable mycelium. So the logs should be freshly cut and then set off the ground to cure until they're ready to be drilled. Once drilled, they should be plugged and sealed with wax to prevent infection from undesirable mycelium. The logs are then set in a moist, sun dappled area of the forest to incubate.

It's a lot of work, but it can easily be broken into steps just right for a production line method, similar to our juicing party. When we assembled I noticed Greg was dressed in throw-away sweats with a boonie hat pulled low, just in case the party degenerated. Silly, since we were too mature to let that happen twice . . . and there's no fun in pitching mushrooms.

We settled into the routine of the work and conversation followed. After our previous, superficial bonding over apple mash and a life rescued from death, it was time to go deeper. Time to confess. I announced we were selling the farm. Greg reviewed our long litany of failures that lead to the inevitable conclusion. I noted we had already engaged a realtor. Kathy and Mark were shocked, then disappointed. We had been off to such a good start as neighbors and now this!

It did feel like a betrayal, and I suppose withholding the information is a deceit by omission; but that's what happens when you're in transition. It's hard to be authentic when you're standing between two realities. We were farmers, but failed farmers who might not be farmers much longer but secretly still wanted to be farmers. Well, the deluded part of us still wanted to be farmers—the part that was drilling holes in mushroom logs we probably wouldn't be around to harvest.

The Sunketts, being new to the lifestyle, tried to defend our delusions. This was good, because it's easier to see the delusions when someone else is saying them.

"Why?" they asked. "You have such a beautiful farm, why would you give this up?"

"Because it's incredibly hard work for low pay and high risk. Better to ask, why would anyone want to keep it?"

"High risk?"

"Yeah, farms are expensive. All that capital tied up in land and machinery and all of it bet on this year's harvest. Our entire life's savings depends on the right amount of rain fall.

"And then there's the risk to life. I almost died. I didn't, but only thanks to you. Greg had two close calls this year. If it's not a tractor rolling over or a tree bucking back when falling it, then it's a hoof to the head or a hay elevator catching a loose sleeve and ripping an arm from a socket. It's risky in every sense of the word."

"It's risky driving the freeway to work. And depressing, too. At least it's pretty here."

"So what? I can live in the city and drive to the pretty on weekends." I was digging in.

"Yes, but would you have that connection to the land? To this lifestyle?"

"Right, the hard-work, low-pay, high-risk lifestyle? That 'back to nature' thing is a romantic delusion. You're gardening, so you've minimized the risk and that makes sense. But, small farming is doomed to fail. The economic winds are always blowing against us. It's a small boat on a very choppy sea and at some point you're going to swamp. And the thing about being in a small boat—there's no dinghy."

Well, that shut the conversation down. Greg glowered at me, but I was being authentic. Or, at least, I was representing truths I had tried hard not to see before. Everyone became very focused on their individual tasks. Until Mark put down his drill. He had that far away gaze of someone churning an idea in their head.

"I think it has something to do with scale. That's what makes the difference between city and country life."

"Scale? Really?" Greg was issuing the challenge. I decided, better if I just kept working.

"Yeah, it's this weird, reversal thing. Cities are these big, electric landscapes created by people. We're at the very center of city life. We build the buildings, turn on the lights, create the jobs—and yet everyone in the city feels small and insignificant. We go home to sleep in our little boxes, watch movies through our little boxes, text through our little boxes. It's all kind of small."

"Small? I would have said self-absorbed."

"Yeah, kind of the same thing. When we make ourselves the center of the universe, we make the universe our size—small."

"Okay, maybe."

"Look at our fascination with celebrities. What are we at now, the D-list of celebrity status? Why do we care so much what some celebrity is doing? We feel small and imagine that being a celebrity would make us feel big and worth knowing."

"Right. And now we can all be stars on YouTube. With our own PR firms on Facebook." Kathy had caught Mark's far-away gaze. Our production line was breaking down.

"Yeah, except the more cyber-connected I am the more cyber-needy I feel. Like that saying, 'I don't know I've had an experience until I see it on Instagram.'"

"And I don't know how I feel about it until I see the number of 'likes.'" Kathy added.

"Exactly. But out here, nature is big and people are small. I'm just a small part of it . . . but I'm part of it. Being small, in nature, feels 'right-sized,' instead of insignificant."

"That's kind of a contradiction—in the city we're big but insignificant, but in nature we're small but significant?" Greg was still challenging.

"Pretty much." Mark gave a bear-grin, drawing up the corners of his brown eyes with delight.

"As an example, out here, I could care less what some yowser celebrity is doing. I'm more concerned with what my garden is doing.

"And when I go to the Merc, I know Nate is going to call me by name and ask how my garden is doing. I don't feel insignificant and I don't feel big and I don't feel a need to be big or celebrity or whatever. I feel right-sized. It's all about the scale."

"Yeah, and I like what you said about the Merc." Kathy was contributing. This couple took pleasure in stimulating each other's ideas. But then again, they were brightly deluded.

Kathy continued, "There's like this 'see and be seen' thing. At the Merc, people know me by name. I feel 'seen.' I think we only know we exist when others recognize us. When I walk down a city street, no one recognizes me and I begin to feel invisible."

"Right. That's what leads people to wish they were bubble-headed celebs that everyone would see," Mark interrupted.

"Probably," Kathy noted, then resumed. "Here, I feel 'seen.' People know me. And that causes me to acknowledge them—so I 'see' them. And that makes me feel even more 'seen.' Kind of cool it's reciprocal like that."

"So your self-esteem is really based in others' seeing you?" Greg, the psychologist, felt professionally bound to claim this domain by planting the flag of "self-esteem."

"Kind of . . . it's me reflected through the eyes of others. And if there's no reflection, then it's a little like I don't exist. When I walk down a city street, there's no reflection. I'm nothing special."

"Okay, but how special do you feel after putting in twelve hours of back-breaking work?"

Yes, that was me killing the conversation again. But, I was being vigilant against idealizing farm life. To my surprise, it was Greg who answered.

"Maybe not special, but I feel of value. I think we have hard work misconstrued. We talk about it like it's derisive. Hard work is doing something, which is better than doing nothing. That applies to self-esteem too. If you want to feel better, do something. Pick up a shovel. All work is honorable if it feeds my family and contributes to the world."

"I can tell you one thing wrong with hard work. You're too old. And you're getting older, so let's check the romance of 'honorable work.'"

"I can still make a difference."

"Yes, but for how much longer? If you had children who wanted to farm, who could fall in behind you and pick up the slack, then maybe."

"We have a cabin. We could hire farm labor that could pick up the slack."

"Really? And pay them what? We're losing money."

That may seem harsh but that was the reality. We were done as farmers, and no re-visioning of farm life could change the economics. I knew too well the price of those delusions. And Greg knew the price too. He was just succumbing to the sweet allure of the delusion at that moment. The intoxication of good conversation and mushroom mycelium had him lightheaded.

We were all faces down and focused on our chores. The boys drilling, me tapping, Kathy painting the wax. I felt bad about being the killjoy. My mind kept churning those thoughts—the scale of our lives; see and be seen; meaningful work—looking to justify my actions and wondering if they were justified.

This life had changed me in profound ways. It had purged my romantic delusions and replaced them with a sense of purpose—yes, and a sense of place too. After the purge, what remains is the richness of the farm experience itself. It's the richness that occurs when you're doing something meaningful, like growing food, within a context of a people with a specific history and in a specific geography—so it's purpose and place. I can grow food that feeds me, my family, and my community. I can deal with loss because I know there is always renewal. And I know I'm not in charge. Nature is much bigger, and I must make my adjustments. And now, this final adjustment—learning to quit. We were selling the farm.

Those were my thoughts, when Greg blurted out, "Yeah, I guess I'd rather visit a farm than buy one."

I looked up. "Exactly!"

It was a great idea and it had been sitting there, in front of us all along. "We'll rent Annie's cabin."

Greg looked perplexed and maybe a little frightened by my exuberance. "Uh, sweetie, we already talked about this. People are moving out of the country, not into it. No way that rent in rural America will pay the mortgage on the cabin, much less the farm."

"True enough of long-term rental. We'll do vacation rental. People will pay to stay on our farm."

"Why on earth would anybody pay to stay on a farm?"

"You just said it. Same reason people want to buy a farm." I was on a roll.

"People will come for all those romantic ideas about farming, but they will experience the real thing, just in small doses. There's the education about small-scale food production that comes when they pull carrots from the garden. Then there's the stewardship of the land and how that relates to the steam coming from our compost pile. We can share our trails and our creeks and give families a place to reconnect with each other and disconnect from their devices. They'll leave with that sense of scale we've been talking about. It's the perfect antidote to urban alienation. And people will see the work that goes into growing their food, which could lead to better appreciation of the food . . . and of farmers. And we'll be able to pay our bills. Factory farms can't offer this experience; they're too big. Only small, family farms have the 'scale.' It's our niche!"

"No way. I don't want to run a hotel."

"It's not a hotel. It's a farm—our farm. We're just letting them visit in our lives for a little while. It's not a room; it's a particular place, a culture, a lifestyle. You know what it is . . . it's an American heritage. We're preserving a lifestyle and foods that would go obsolete if small farms go out of business. And we're inviting urbanites into our world . . . both as patrons and participants. We're inviting the people we were five years ago."

Okay, maybe not all my romantic delusions have been purged. Kathy and Mark were enthusiastic, but as I noted, they were also brightly deluded. Greg was reticent.

"Do you really think anyone would want to stay on a farm?"

"Do you want to say 'no' without finding out?"

And that's what we did. Kathy and Mark took their plugged logs home and six months later had beautiful Shitake mushrooms for dinner. We opened the gates to our farm and six months later found our niche.

CONSIDERATIONS OF A FARMING LIFE BY THE FARMING WIFE

Umm . . . maybe it wasn't quite that easy. The day after the mushroom party, I placed two phone calls. One to Tanya, the realtor, asking her to put the sale on hold. She probably wasn't pleased, but she was too much the professional to show it. Like I said, I liked Tanya. The other call was to the county to look into zoning requirements. There was no clear precedent, since no one had ever attempted a farm stay, but Oregon's land-use laws were quite restrictive. It took a number of consultations to work it out.

Once we were green-lighted by the county, I designed a web page offering a farm stay to the public. We really had no idea how much disaster we were courting. Visions of small children darting between moving tractor wheels kept my husband up late at night. For me, I felt launching a website was like writing a rescue note, sealing it in a bottle, and throwing it in the ocean.

If you've never heard of a *farm stay*, why would you look for it? Fortunately, people did find my rescue note on the Internet, and parents were not inclined to let their children play beneath the moving wheels of a tractor.

The first visits went quite smoothly. Typically families helped with feeding the animals, collecting eggs, and picking berries and vegetables. The pace was relaxed with ample time for playing in the creek and hiking mountain trails. The children were thrilled with the direct interaction with the animals and reported having "more fun than Disneyland," while parents seemed to be genuinely appreciative of the farm and of the farmers. For us, it was not only financially rewarding, but emotionally affirming as well. We met people from different walks of life and learned about their lifestyles, so it was a more collaborative experience than we expected. Several of them, by asking simple questions about farming, have given us great ideas that have improved our farming practices.

There were several hiccups, so we wrote the book on farm etiquette that I wished I had at the beginning. Basic things like—they're called "rams" for a reason; never turn your back to them if you want to avoid experiencing the reason. And—leave the gate the way you found it (pasture management has some gates open and some closed, depending on rotation).

The biggest area of confusion was the chicken coop. Many people, having never eaten fresh eggs, were alarmed by the bright orange yolk and disposed of them. Others were fearful of finding a chicken fetus when they cracked the egg. They had to be educated on brooding behavior of chickens. Others entered the coop with thongs on their feet, not realizing that, to chickens, toenails look like kernels of corn. And everyone liked gathering eggs for their breakfast—that direct connection to their food. All of these experiences are why the farm stay was a success. It was like nothing these families had experienced before, and they took home wonderful memories.

Not long after launching the website, lightning struck. An editor at *Sunset* magazine found the site and posted a few lines describing this novel vacation idea. That was picked up by a national morning television show, with some help from my state tourism department. I've been booked to capacity ever since. That was years ago.

The first thing families experience when they come to our farm is the quiet. If you live in a city, you probably aren't aware of the urban noise around you. It's the absence of the noise that tells you this is how much noise your brain has been so diligently screening out. Quiet has a way of provoking unique thoughts, reflections, and family connections.

Presently, I'm sitting in the quiet of my farm. Here are my thoughts.

I was not born to be a farmer, but look what I have found.

After five years, our guests continue to enjoy our mushrooms . . . and carrots, and potatoes, and tomatoes, and beans, and onions, and eggs, and lamb. Many return annually for their farm "rescale" therapy. And thanks to them, we are paying our bills. We're not getting rich, and we're still working hard, but that's what we signed up for when we came to the farm. Tater is still tampering with every latch on the farm. I'm still selling the latch that keeps him frustrated and secured. Paco is never without a friend, as small hands brush clots of dirt from his matted hide all summer long. He is a favorite of the children who visit.

And one of those sets of hands belongs to my grandson! With his every stroke I see the hands of a farmer developing a feel for livestock and the land. Maybe I just needed to wait a generation to get my legacy. Of course, like my daughters, he may decline the farming life. But for now I'm hoping he will choose it. And yes, that's the same romantic, agrarian delusion that brought me here, but, to paraphrase a friend, some delusions are worth doing again.

I'm glad Paco is a favorite with the children because I know that would have pleased Jack. When I see Paco, I wonder if Jack isn't there in some form directing each child's hand and smiling his mischievous smile. He is reminding me that it takes a love of folly to survive the country. Knowing that makes me smile.

And while feeding the sheep, I can gaze up in the woods where I know a grandfather tree holds the memories of all my loved ones. With every gust of wind, I know they're dancing. Their memories are not lost to me. And as I gaze, I feel the scale of this life and the passage of its seasons—reminding me, relationships are to be cherished.

And every morning I go walking with Sarah and my new friends, Kathy and Mark, and a horde of happy dogs. Sometimes we pass Brick, with his fishing pole and big smile, off to his mayoral duties. And Nate at the Merc, and Will on his tractor, and even Luther on the porch of his bunkhouse smoking his pipe of an evening. I see them all and I'm seen by them. Knowing that makes me feel connected to this community, to these folk, and their ways.

And every evening, I lie next to an old farmer who loves me. In the darkness, I rest against the warmth of his bent frame, content in the knowledge we are a support unto each other. Yes, I am still here in this place with my stubborn, farmer's heart. And I am grateful, so grateful, for this life.

EPILOGUE

STAY ON THE FARM

We stayed on the farm because we opened a farm stay. We also realized our story was not unique. If 12 percent of farmers are producing 90 percent of the food, you can bet the other 88 percent, like us, are not profitable. These are the small farmers, but given that they're 88 percent of the farming population, can we really call them small? Many are farming marginal lands that big ag would not deem worth the effort. Considering also that 80 percent of them are working other jobs, it should be clear that this is a threatened lifestyle.

And rightfully so, argues the economist. After all, how many of us turn to the village cooper when our wagon wheel breaks? Some jobs are slated for obsolescence, and profitability is society's way of winnowing the chaff. So, are we now willing to hand over our food supply future entirely to big ag's industrial style farming? They are quite willing to accept that contract in exchange for plentiful and cheap food. If not, what are we willing to do? Support increased taxes that provide subsidies for small, inefficient farms? Pay more for our food? Based on current trends, the answer seems to be, "yes, but only a little bit more." There is a definite limit to how much more the public will pay for organic food, either in the marketplace or through federal subsidies.

This is why farm stays offer such an innovative third option. They provide a service to the public that big ag not only can't but wouldn't even want to provide. And they support an American heritage, the small family farm, without raising taxes or food prices. They also provide an experience for the public that creates memories. That's why a farm stay is not just a bed & breakfast on a farm. It's not a room, it's an experience. It's a window into the rural lifestyle. Think of it as the easiest trip into a foreign country you'll ever take.

When we realized our farm stay discovery offered a solution to a national problem, we decided to organize a national nonprofit association to connect farms with the touring public (www.farmstayus.com). We quickly found that we had invented a wheel that the rest of the world was already using. Seems the Brits have been offering farm stays since the early 1970s. In France they're called *gites*, and in Germany, *schlaf im stroh* (sleep in the hay). In the 1990s, Italy became concerned about the influx of the rural population into cities to escape rural poverty. They countered with an aggressive *agritourismo* policy that turned farm stays into a multibillion-dollar industry that is effectively reversing the rural to urban influx.

So, like us, if you've never heard of a farm stay, you're probably an American. Strange that the greatest agricultural power in the world doesn't have farm stays. Or perhaps it's because of the American trend toward industrial farming, with its heightened efficiencies, that we don't have farm stays. Doubtful anyone wants to spend their vacation in a chaise longue on a feedlot. You have to have traditional farms to have farm stays. If the trend is toward big ag, it raises the question of what we as a culture are in the process of losing—without quite knowing it's being lost—and what we might be willing to do about it.

What is the experience offered by a farm stay? The answer is as broad and unique as the farms that host the stay. Accommodations can range from tents to treehouses to posh digs. Most stays allow some involvement in farm life, but guests are never required to work. Heritage farms tend to be diversified operations, usually involving a mix of livestock and produce. The addition of livestock is important because interacting with animals is often the bridge that allows people to connect to the rural experience. Most heritage farmers have a strong sense of place and tradition and acutely feel the threat to their lifestyle, so they want to share their folk knowledge. That said, farm stays are not usually heavily time-structured, allowing ample freedom to explore beyond the farm or to kick back with a good book by a rippling stream.

Why should small, inefficient farms enjoy public support? By staying at a farm, aren't we just prolonging their inevitable demise? Might it be better to prohibit farm stays from operating on land designated for farm use? These are questions I hear routinely from a vocal minority in the big ag farm lobby. There are a dozen rebuttals, but the simplest is, mother nature loves diversity more than our industrial economy loves efficiency. It is wise to remember that the Irish experiment in monoculture farming resulted in a potato famine, which went on for a decade and killed millions. It was a problem the inefficient Andean farmers, who developed the potato, did not have because they farmed multiple varieties. Small farms may be inefficient, but they are also diverse, and we should not rush to close the door on the options they provide.

This is especially true at this juncture of our history. The organic movement has been percolating since the 1930s and seems to be gathering momentum with a new generation of young farmers who combine a sense of pragmatism with a valuing of sustainability over quick profit. These young farmers are networked and sharing results as they engage in novel experiments in agriculture, economics, and social order. You can meet many of them at your local farmer's market. They are happy to share their experiences and visions. It is too early to predict how these experiments will shake out, but it seems likely they will be a source of innovation for intensive, stratified farming of small land parcels. We need to preserve our small farms so this new breed of farmer has a platform for their innovation.

So, that's my pitch for farm stays. Hopefully you can see it has mutual benefits for both urbanite and ruralite. What you can't see is the ripple effect it has in the local rural economies that are presently collapsing in on themselves like black holes. If you choose this vacation option, rest assured your farm stay helps a heritage farmer stay on her farm.

ADDENDUM: A PRIMER ON FARMING

ECONOMICS: JUST THE FACTS

The United States is blessed with more arable land than any nation on earth, at present. America also has the most productive farmers in the history of the world. These joint blessings allow Americans to enjoy the most affordable food of any developed nation. And that in turn provides Americans more discretionary uses for their excess income. What follows is a brief synopsis of the economics of farming. Think of it as a health report on your food supply.

Much of the data being reported is based on the 2012 USDA census. The census was published in 2014, but the data was collected two years earlier. Census data is collected every five years by the USDA and made available through its website. Note that a farm, as defined by the USDA, are any operation that contributes a minimum of $1,000 per year of agricultural product to the economy.

Other sources of data include the scholarly works of David Danborn, *Born In The Country*; Paul Conkin, *A*

Revolutiojn Down On the Farm; and Daniel Imhoff, *Food Fight*.

PRODUCTIVITY

The US population is approximately 313,000,000, of which 1 percent (3,180,074) are farmers. There are 2,109,303 farms accounting for 914,527,657 acres in production with total ag sales of $402,697,828,000. This figure includes $8,053,346,000 in government subsidies.

Farms with total ag sales equaling $25,000 or less account for 67 percent of the total number of farms and 2 percent of total sales. Farms with ag sales between $25,000 and $1,000,000 account for 29 percent of total number of farms and 32 percent of total sales. Farms with ag sales above one million dollars account for 4 percent of the total number of farms and 66 percent of total ag sales. Clearly a small amount of farmers are responsible for a large amount of US food production. To see this more simply, if we use $250,000 in ag sales as the dividing line, then 88 percent of farmers are below that. The 12 percent of farmers above that line are responsible for 89 percent of total ag production. Given that consolidation is ongoing, that number could easily be 10 percent of farmers responsible for 90 percent of food production by the next census. That would be 0.1 percent of the population responsible for 90 percent of the food produced in the United States.

PRICING

The increased efficiency on the farm shows up in the supermarket. In 1945, 36.4 percent of family income went to food purchases. By 1970, it had dropped to 23.2 percent of family income, and by 2002 it was 15 percent. The price of food is getting cheaper, which is good for consumers, but it also means the margin of profit for farmers is getting narrower. As another example, from 1945 to 1970 the price of farm produce rose 41 percent while the broader Consumer Price Index rose by 116 percent. Food is becoming more affordable.

PROFIT

In 2002, 46 percent of farms made a profit. Of those making a profit, the average income was $56,000 per year, including subsidies. In 2012 the percentage of farmers turning a profit had not changed, but the average income was dropping. In 2002 the average income from a mid-sized farm (between $100,000 and $250,000 net production) was $30,000 per year, with 50 percent of that coming from subsidies. By 2007 it had dropped to $26,000 per year. Most families can't live on this amount. Currently 80 percent of farmers take off-farm jobs to supplement their income. Additionally, 53 percent of farm couples are both working off the farm.

Since 1910 farm revenue has increased six times while farm overhead has increased sixteen times. In 1945 the average farm returned 17.5 percent of its value in income. By 1970 that figure had dropped to 6.4 percent. Today roughly twelve cents of every dollar spent on food returns to the farmer.

FARM SIZE

Essentially, all arable land in the United States is under production. There is no significant amount of new land to be developed. Total acres under production are declining as cities are expanding. In 1982 total acreage under production was 986,796,579, compared to 914,527,657 in 2012. While there is a decrease, it appears to be declining at a modest rate. Still, it deserves our concern, especially if improvements in food production are not able to keep pace with the decline in acreage.

Average farm size expanded from 145 acres in 1920 to 434 acres in 2012. Average farm size is probably only a rough gauge of the trend toward consolidation because very small "hobby" farms are abundant, skewing the average. From 2011 to 2012 farms producing over $500,000 in ag sales expanded by 8.6 percent in number and 3.7 percent in size (i.e., added acreage), while farms producing less than $10,000 decreased by 2.5 percent in number. Overall there is a broad, long-term trend towards consolidation. As profit margins narrow, there is increasing pressure to increase acreage.

DEMOGRAPHICS

A very real concern in farm policy is the aging of farmers. Over the past thirty-five years, the USDA has reported the average age of farmers to be increasing steadily. From 2007 to 2012 it rose 2 percent to 58.3 years. Also, the rate of new farmers (farming less than ten years) has decreased by 20 percent over that same time frame. Taken together, the data points to a rapid increase in start-up costs in farming, making it cost-prohibitive for young people to become farmers. This could lead to a serious skill gap when the older generation retires. The problem is very troubling with no clear answer yet.

Gender roles have witnessed considerable change on the farm. Women constitute 30 percent of all farmers, with 14 percent being primary farm operators. Most of these are small farms, with 91 percent of female-operated farms having less than $50,000 in sales. It's too early to say what this influx of women running small farms will mean to our food culture, but I suspect it will be significant.

Another area of change is the increasing diversity among farmers. In the past, barriers to minority farmers have proven to be a disservice to the economy and the moral fabric of the country. Those barriers and inequities are now lifting, allowing for diversity that may give rise to more innovation. However, farming remains a predominantly white, male occupation.

HISTORY

Our nation was created from farming. The English colonized America in large part for the purpose of farming. The values of independence and self-sufficiency required of farming gave rise to the notion of self-governance. Our founding fathers—Washington, Adams, Jefferson, Madison, and Monroe—were all farmers. Jefferson in particular made agrarianism a centerpiece of his policies and politics. Perhaps because the nation was founded by farmers, our government is uniquely sensitive to the needs of the agricultural sector both as an economic driver and as a cultural institution for the preservation of core ideals. Thus, any discussion of American farming requires some review of seminal moments in US history.

American agriculture got its first big boost from the Congress of 1862. With the exit of Southern Democrats, the Republicans were free to legislate broad federal policies. They created the Department of Agriculture, the Homestead

Act, the Merrill Land Grant College Act, and the Pacific Railroad Act. The benefits to agriculture from the first two were obvious, but the second two were just as important. The Land Grant College Act encouraged the development of state colleges specifically for the study of agriculture, greatly advancing new technologies in agronomy. Today, our state universities are engines of innovation across a wide variety of disciplines including agriculture. The Railroad Act was intended to connect the two coasts of our country, but it also opened the fertile Midwest to farming and the shipment of agricultural produce. Taken together, these four acts were visionary in their understanding and anticipation of the role of agriculture in the wealth of a nation.

By the 1880s, America was reaping the rewards of its investment in agriculture, but it was turning into too much of a good thing. While agricultural abundance was good for the nation, it was bad for the farmer. Flooded markets destroyed farm prices, leading to bankruptcies that would roil local economies. For the next fifty years, the nation would sway between violent boom-and-bust cycles as it searched for a way to stabilize commodity markets.

The great boom years were 1910–1920. American innovation pushed production to new levels just as World War I shut down European agriculture. The period just before the war (1910–1914) is considered "ideal" in that it represented peak pricing without inflation from war. It's the standard the government uses to this day to establish "parity" in agricultural pricing. The year 1914 also saw passage of the Smith Lever Act establishing the Agricultural Extension services, essentially capturing the developments from college ag schools and then assisting farmers in rapid implementation of best farming and conservation practices.

The bust came in 1920 when European ag came back online. The return of competitive pricing caught many American farmers over-extended, leading to bankruptcy, foreclosures, and a particularly severe contraction in rural economies. Just as the rural economy was starting to regain momentum, Wall Street collapsed (1929), again destroying capital and credit. Then came several years (1932–1937) of severe drought in the Western plains. Together, these events kept farmers on their backs for the better part of two decades.

Farmers sought aid from the federal government, but Republican administrations were adverse to direct intervention. Instead, they encouraged farmers to organize into cooperatives to gain more leverage on the market and stabilize commodity prices. Unfortunately, most of the cooperatives were either too broad in scope or attempted to package groups together that were too disparate in their goals to be effective.

Federal activism came in 1933 with Roosevelt's New Deal. The farm policy focused on three parts: 1.) Restoring credit by lowering mortgages and extending the term of the loan (later adopted by the FHA), 2.) Paying farmers a subsidy to limit production, and 3.) Offering farmers a crop loan, creating a bottom below which prices wouldn't go. The crop loan meant the government would buy the crop if the market dropped below that predetermined amount. However, once the government bought the crop it couldn't sell it without tanking the market. So the government created granaries that became a source of emergency food relief and that eventually grew into the food stamp program of today. Roosevelt also initiated a rural electrification project (beginning with TVA) to bring electricity to every farm. It took thirty years to complete but transformed farm life and greatly increased productivity.

The goal of the New Deal's farm policy was to strengthen the nation's agricultural sector, especially by modulating the boom-bust cycles. It was always geared toward big ag for the simple reason that that's where the biggest effect on the ag economy can be achieved. It was marketed as helping poor farmers, and it did. But its emphasis has always been with big ag, and big ag benefitted disproportionately. Given the goal, that is a strategically smart approach to farm policy.

These three interventions were intended to be temporary until the country came out of the economic crisis. But the Great Depression gave way to World War II, and the country still needed its agricultural sector performing, so the supports were extended. Then came the rebuilding of Europe and the Korean War, and eventually the New Deal became an accepted part of Farm Bill policy.

THE FARM BILL

Essentially, the goal of the Farm Bill is to ensure a reliable and safe food supply for the nation. It has typically been divided into three parts: supplemental nutrition (about 70 percent of the funding); income/price supports for commodity crops (about 20 percent of the funding); and conservation incentives (about 5 percent of the funding). Supplemental nutrition tends to be a concern for the more urban states while the price supports are a concern for rural states. Subsidies have always been a hot button issue, so let's look at a few of the facts before entering a discussion.

Typically, farm subsidies account for about 3–5 percent of the federal budget. Sixty percent of all farmers receive no subsidies. The top 1 percent of all farms received 20 percent of the subsidies. The top 15 percent receive 50 percent. Ten states received 66 percent of the subsidies yet produced only 33 percent of ag GDP. California, tops in ag production of all the states, received 10 percent of subsidies. Cotton has, by far, been the most subsidized of all crops.

Clearly there have been inequities in the distribution of subsidies, which adds to the controversy. Beginning in 1996, Farm Bills have increasingly attempted to eliminate subsidies and replace them with crop insurance. With crop insurance, the farmer receives payments for losses due either to nature or adverse market conditions. The 2014 farm bill has eliminated direct payment subsides (i.e., paying farmers not to farm), but some critics feel various crop insurances look similar to the direct payment subsidies in the way they are structured.

Farm Bills are some of the most complicated pieces of policymaking our government engages in, with very big consequences on the economic health of the nation. As citizens, we want to pay attention to this legislation as it's being enacted.

THE GREEN REVOLUTION

The Green Revolution is a term generally referring to a dramatic increase in agricultural productivity that occurred from

1950 to 1970. According to David Danborn, the increase was the result of technological innovation across three broad areas: plant and livestock breeding; mechanization; and synthetic chemicals. While these innovations peaked in the second half of the century, they begin at the turn of the century and continue to this day.

The term Green Revolution originated as a geopolitical manifesto to develop plant varieties that would assist Third World countries in achieving agricultural self-sufficiency. The work of Norman Borlaug, winner of the 1970 Nobel Peace Prize, is a prime example of this use of the term. Beginning in the 1940s, Borlaug developed wheat varieties in Mexico that would raise the nutritional standards of millions of starving farmers in Africa and India. However, his wheat also improved production in First World countries, leading to the broadening of the term Green Revolution to mean any significant increase in agriculture as a result of a scientific approach.

For corn, this began in 1926 when Henry Wallace developed the first closed pollinated corn hybrid that increased corn production by 20 percent. To stay competitive, farmers had to buy the hybrid, and they had to return every year to buy it since it can't reproduce naturally. Thus began an industry that led to GMOs and seed and gene patents. Livestock followed a similar pattern, most notably with artificial insemination, which greatly increased the advantages of selective breeding. Milk production doubled in the second half of the twentieth century due to selective breeding. This has had profound implications as concerns arise about intellectual property rights and the long-term consequences of injecting synthetic or altered substances into natural systems.

Mechanization also saw dramatic increases in the latter part of the twentieth century. While cars were well established in the American landscape by 1920, tractors did not find acceptance. Largely, this was because pull-behind farming implements had to be raised and lowered at every turn in the field, proving unwieldy compared to horses, and they were tied to the speed of the tractor. In 1926, the addition of the power take-off (PTO) to the tractor allowed machinery to operate at the speed of the engine rather than the speed of the tractor, greatly improving the utility of tractors. Unfortunately, the improvement came as farmers were facing severe capital restrictions from the Depression, delaying the roll out of tractors. In 1939, there were still only two tractors for every nine farmers. By 1970, there were 1.6 tractors for every farmer. Perhaps the greatest leap forward in labor-saving machinery was the self-propelled combine, introduced in 1948.

The down-side to mechanization is that farmers became increasingly dependent on fossil fuels to the point that fossil fuels are now essential for the nation's food supply. Fossil fuel is not only critical for mechanization, it also is vital to the third area: synthetic chemicals. Of the three nutrients necessary for plant production (phosphate, potassium, and nitrogen), nitrogen is most critical and most easily leached from soils by rain, meaning it needs to be supplemented. In 1909 in Germany, Fritz Haber produced the first synthetic nitrogen using methane derived from natural gas. In 1943, America began its first large scale commercial production of ammonia nitrate in Muscle Shoals. Today, synthetic fertilizers are abundant and cheap, freeing the farmer from the slow composting process in organic fertilizer. This has greatly increased crop yields, but at a cost to the environment, upsetting the pH of fragile ecosystems.

Another production benefit of the synthetic chemical industry has been the introduction of potent herbicides and pesticides. Insecticides were introduced in WWII to relieve soldiers of irritation from lice and mosquitoes. They were used heavily on farms after the war, resulting in considerable health and environmental damage. Some insecticides have simply been eliminated as too detrimental, but others remain an indispensable part of high-production farming, which attempts to manage the risks these poisons always pose. Herbicides are even more necessary for their labor-saving advantages, but they are just as problematic and their use should always be judicious, if at all.

Finally, the use of antibiotics and steroids has led to some of the most dramatic improvements in production from the chemical industry. They may yet prove to be some of the most destructive as well. They are currently under intense scrutiny from the scientific community, but initial findings suggest extreme caution, if not total elimination, of their prophylactic use in farming.

COMMUNITY SUPPORTED AGRICULTURE (CSA) AND THE ORGANIC MOVEMENT

A CSA is a group of subscribers who pay an up-front fee to share in a farmer's harvest, usually of produce. In its simplest form, it is seed money and crop insurance for small farmers. The farmer receives money for the planting and is guaranteed that a certain percentage of the crop will sell. The subscribers are sharing in the risk of farming, but are likely to receive good value for their investment, if for no other reason than the elimination of a middleman. As noted, this is the simplest form of a CSA. It is a very flexible construct that is limited only by the creativity and cooperation of the people committing to the contract. For example, the investors may exercise varying levels of control over what gets planted, how it will be farmed, who is doing the farming, and where the farming takes place.

CSAs owe their legacy to the "biodynamic agriculture" philosophy of Rudolf Steiner. In the 1920s, Steiner introduced notions of ecology and soil sustainability that eschewed the use of synthetic fertilizers and pesticides. He viewed the farm as an "organism" within a natural context—a fully self-sufficient system within nature. He also incorporated elements of mysticism, such as astrologically guided planting. The term "organic" is credited to Lord Northborne, who used it to refer to farming practices that use Steiner's farm-as-organism approach. However, Sir Albert Howard, with his application of scientific methodology in the 1930s, gave organic farming its first big push forward.

Jan Vander Tuin, a proponent of biodynamic agriculture, created one of the first CSAs on a farm in Massachusetts in 1986. That same year, Anthony Graham and friends created a CSA in New Hampshire, launching the CSA movement. CSAs continue to evolve, challenging traditional models of economics and private property in producing food and culture.

SUMMARY

America is doubly blessed by having the largest amount of arable land of any country and the most efficient farmers in history. This results in Americans enjoying the lowest food prices of any industrialized nation and the highest standards of living in the history of the world. I know that sounds inflated, but consider: the average American lives better than most kings have throughout most of history. Just look in your refrigerator if you want proof.

That efficiency is the result of a very competitive business model, and anyone considering farming as an occupation should be aware of these facts. Farming is a high-capitalization, low-profit, high-risk business that depends on large scale to achieve profitability. Because of the risk, it is in the interest of all Americans to use government support to stabilize commodity markets and reduce that risk. Without supports, shocks in the commodity markets ripple through the broader economy—meaning everyone suffers. Of course, finding the right balance of government support without creating dependency is the reason Farm Bills are so contentious.

There is a long-term trend toward consolidation in which big farms are getting bigger to survive. This also means there are fewer farmers on the land, meaning rural communities are getting smaller and more isolated. Increasingly, mid-size farmers are being forced to take second jobs off the farm to stay viable. Small farmers have always had jobs off the farm.

The high cost of farming creates a variety of social problems. For farmers, it creates enormous problems with legacy. If farms need to get bigger to survive, then it isn't a viable practice to break them up to give to your children. Further, since they are getting more efficient, it means the family farm can support fewer people in the family. The bigger and more expensive farms get, the harder it is for young people to afford the buy in, meaning it can take a lifetime to afford the family farm. All of that makes it hard for farmers to achieve equity with their children when they want to retire.

Shrinking rural communities also makes it less desirable to live in the country. Fewer people on the farm results in a negative spiral of less money in the community, which leads to fewer resources and less opportunity. It also means less political clout to get anything accomplished.

Recognizing these economic realities, there are lots of reasons to choose to live in the country—many of which I've mentioned in my farmoir. What is clear is that if you are not going to run a large-scale farm operation, you need to find that niche. The Internet is a vital tool for finding markets and testing niche agriculture. It may prove as necessary to the next generation of farmers as electrification was to the previous generation. To my continued amazement, young farmers are using old farm technology in novel ways. Guided by a vision of renewable farming, they are transforming the rural landscape, producing new models of food production and reinvigorating rural communities. None of my young acquaintances are on a path to monetary wealth, but they are certainly finding a richness of life and a sustainable way of living.

I still occasionally encounter that subset of big ag farmers, espousing their Ayn Randian view that only big farmers should be allowed in the sandbox. According to them, only big ag is efficient, and efficiency is the only metric that matters. They tend to be most vociferous around political issues where they perceive threats to their subsidies and too often they use their wealth to stifle discussion. Thankfully, they are a minority of the big ag farmers. I think there's room for both big ag with its efficiency and small ag with its choices, but it will take support from an informed public to ensure the survival of small ag and the preservation of heritage farming. If you agree, I hope you will look for opportunities to support small ag through farmer's markets, CSAs, and, of course, farm stays.

ACKNOWLEDGMENTS

Besides family, friends, and neighbors who supported us on the farm, there have been several people of note who made this book possible. Foremost is our agent, Christopher Rhodes, who believed in this story from the beginning, providing us with guidance and support when it was most needed. Also, our editor, Maxim Brown, with Skyhorse Publishing, whose patience and critical insights converted a rough manuscript to a finished book. Other notables include Kristi Crawford, for the wonderful cover photo; Sonya James, who graciously modeled in our barn's hayloft on a blustery January day; Paul Deatherage and Joanna Lezak for sharing their personal photos; Kate and Dennis Rivera for providing a visual diary of our farm; and Justin and Teagan Moran for shepherding our farm into a more organic future.